MEASURING METABOLIC RATES

MEASURING METABOLIC RATES

A Manual for Scientists

Second Edition

JOHN R. B. LIGHTON

OXFORD
UNIVERSITY PRESS

OXFORD
UNIVERSITY PRESS

Great Clarendon Street, Oxford, OX2 6DP,
United Kingdom

Oxford University Press is a department of the University of Oxford.
It furthers the University's objective of excellence in research, scholarship,
and education by publishing worldwide. Oxford is a registered trade mark of
Oxford University Press in the UK and in certain other countries

First Edition published in 2008
Second Edition published in 2019

Impression: 1

Published in the United States of America by Oxford University Press
198 Madison Avenue, New York, NY 10016, United States of America

British Library Cataloguing in Publication Data
Data available

Library of Congress Control Number: 2018959188

ISBN 978–0–19–883039–9

DOI: 10.1093/oso/9780198830399.001.0001

Printed in Great Britain by
Bell & Bain Ltd., Glasgow

PREFACE TO THE SECOND EDITION

Publishing the first edition of this work was a most interesting exercise, and I learned much from the talented crew at Oxford University Press. They shaved my rough prose, dressed it in a suit, and made sure that the pocket handkerchief in its breast pocket was precisely square-folded. I owe them a debt of gratitude.

Metabolic measurement is a common requirement in science, medicine, and education, and pitifully few people know how to do it well. The first edition was well received and addressed the real need for an advanced handbook in this area that combined practical details with enough theoretical background to satisfy people interested in the why behind the how.

Interestingly, pirate copies of my book in .pdf form are very popular. Berry Pinshow of Ben Gurion University was kind enough to send me one and I was impressed by its high technical quality. Berry told me that it was much sought after by his students, and I find myself using the copy he sent me more than my own paper copy of the book because of the ease of electronic searching. Copies are available on dozens of torrent sites.

Then my book's errata started piling up, thanks to input from astute or simply nitpicking (bless 'em) readers around the world. My web site, respirometry.org, lists the errata, but I'd rather have corrected the errors in a new edition. Anyone who has wasted hours before finally figuring out that an equation is wrong and then located the erratum after a long search will understand. Several plumbing diagrams needed significant revision in the light of experience. Plus, in the 12 years since writing the first edition I've learned a lot—including the practical aspects of direct compensation for water vapor dilution, which can eliminate the need for awkward chemical and thermal water vapor scrubbers. I've also become much more engaged in two areas of metabolic measurement that are exciting, challenging, and relevant to human welfare: metabolic phenotyping of research animals and room calorimetry of humans. Chapters about these areas would, I thought, widen the range of readers interested in the book—especially because a thick mist of myth and confusion clings to each of these fields.

So, I went back to Oxford University Press and suggested that the time was right to consider a second edition of my book with corrected errata, several significant revisions, and two new chapters. After an initial rejection by the USA OUP, I was about to publish the book on Kindle when the visionary Ian Sherman from the Oxford office of OUP decided to invite a revised edition. And here it is.

Why do I do this? Not for fame or wealth. Rather, I deal daily with people who need advice on making metabolic measurements and I want them to succeed. Why? Because in my opinion, science is the human species' apex achievement, one that may one day allow us—perhaps the only intelligent life in our galaxy—to reach the stars. Science is fueled by successful research, and a researcher's success in the field of metabolic measurement is best served by a book with expanded coverage and the correction of annoying but inevitable first-edition errata.

Hopefully this guide will also be available as an eBook. Of course, the advantages of the eBook format are many. Portability, accessibility, and searchability are chief among them. Apart from being more up to date, an eBook also offers a subtle improvement over even the "official" pirated copy of the book's first edition, which sports typographical fossils such as the ligature of two graphemes into a glyph, for example "fi" and "fl," that frustrate electronic searching (try searching for "Nafion" or "flow rate" in the pirate copy; you won't find them in the text).

I make no secret of the fact that I am a recovering academic, though still active in research; and president and CTO of a company, Sable Systems International, that makes metabolic measurement equipment and systems primarily for scientific research. That said, I maintain an impartial attitude. Some readers have remarked that I'm crazy to write books that tell competitors and sundry imitators how to make excellent metabolic measurements. But they miss the point. I don't care about competitors, let alone imitators. This book is directed not to the makers but to the users of metabolic measurement systems. Making excellent measurements requires excellent equipment, and the art and science of making such equipment is exceedingly difficult. When it comes to that area, which this book does not cover, I have much to teach competitors and imitators, but they won't learn it here. My aim is, rather, to help end-users make excellent measurements that add to the knowledge-base of our species.

Although I don't cover equipment manufacture, and (as pointed out in the first edition preface) don't have the space to include a guide to practical electronics and so on, a welcome tectonic shift in the approachability of electronics and microcontrollers appropriate to laboratory settings has occurred in the last decade. Acquiring skills in these areas is a huge scientific advantage. For interested readers, I have added a brief Chapter 20 with some guidelines, sources, and recommendations.

Once again, I acknowledge and thank my friend, partner, and CEO Robbin Turner for her encouragement and negotiating skills, and the Helen Riaboff Whiteley Center retreat at the Friday Harbor Laboratories for making this book possible via a brief shelter from the whirlwind of my usual life. Thanks, too, go to Ian Sherman for encouraging me to return to the OUP fold after my dalliance with Kindle (nice though Kindle is).

I hope you will find this book useful. Supplementary material appears in the companion web site, www.respirometry.org, at which I can be contacted (if that doesn't work, try lighton@sablesys.com). My seditious and opinionated Twitter feed is available @sablesys. Constructive suggestions, error reports, poems, and pressed flowers are always welcome.

PREFACE TO THE FIRST EDITION

This book is for practicing scientists or future scientists who need background information and application specifics to develop their skills in research-grade respirometry. The emphasis is on measuring the metabolic rates of diverse creatures in air; aquatic measurements are covered, but in less detail. My background as a comparative, terrestrial animal biologist means that the text is biased somewhat (well, let's be honest, considerably) toward that discipline. That's not a bad thing because comparative biologists may, over the course of a few years, work on bacteria, flies, bandicoots, fish, hummingbirds, and killer whales. They go where their questions take them. Because they are exposed to a wide variety of measurement challenges, they have learned, or been forced against their will, to be resourceful. That said, people from other disciplines (genetics, pharmacology, microbiology, biochemistry, etc.) will also benefit from this book because the metabolic measurement principles discussed are universal. (I won't say anything about the botanists. They live in their own world.)

It is possible to measure metabolic rates without understanding what you are doing. In doing so you may think, or hope, that the data you acquire are accurate. In fact, this approach is pretty much the rule, especially in the time-pressed biomedical community. Anyone who has purchased this book in the hope that they can metaphorically throw it at a gas analyzer and without thought, effort, or understanding produce numbers will succeed. As for what the numbers will mean, that may be another story. Thus, it is possible to abuse this book and apply it without understanding, but my hope is that the text will discourage this approach. I suggest that you first read this book as an idiosyncratic sort of novel, including the parts you know you'll never need, and then with that background return to the sections most relevant to your scientific goals. If you don't have the time, you should pass it on to a factotum who will do your understanding for you. Don't feel slighted by this judgment; you may go far if it applies to you. As Villiers de l'Isle-Adam's Axël might have said, "Understand? Our servants will do that for us." The goal is for *someone* in the operation to understand. If you are that person, welcome: This is your book.

I assume that the reader has a science background and is familiar with life science basics such as the definition of basal metabolic rate, the thermal neutral zone, and so on. To keep this book to a manageable size, its emphasis is on understanding the how to. The *why* behind the how to is your department. Nevertheless, Chapter 15 covers some aspects of the analysis and presentation of metabolic data that deserve to be better known.

Much of this book, when applied in practice, will best benefit you if you have some experience in computer-assisted data acquisition and analysis. Some areas in which off-the-shelf solutions do not exist may require experience in electronic and/or mechanical fabrication, about which I wish I could give guidance in this book. I can't; there isn't enough room. I wish I could recommend a good book to get you started in these areas, but with the qualified exception of Horowitz and Hill's *The Art of Electronics* (Cambridge: Harvard University Press, 2015), none exists.

I was forced by the nature of my own research to develop skills in electronics, fabrication, and computerized data acquisition and analysis. The process took many years and was intensely frustrating. One favorable spinoff was my ability to say sayonara to full-time work at the academy a decade after co-founding the company of which I'm currently president, Sable Systems International. My research, scientific collaborations, and teaching continue unabated, but on my terms. Authoring this book marks, as of 2008, 25 years of experience in animal respirometry, and brings that experience into focus from the joint perspective of a user and a developer of techniques and instrumentation in the field. Having helped many beginners and senior scientists (including Nobel prize winners) get started, I know from experience that this joint perspective is sorely needed in a field as demanding of its users as respirometry. But one should never lose sight of the fact that it is the living, breathing researcher with knowledge and insight who dictates the success or failure of a project. Tools help but are only that—tools, but like all tools, they should be understood. This book will help the researcher to make the best use of them and guide him or her to the types best suited to the intended application.

Many bits and pieces are required in a respirometry laboratory, ranging from tubing, tubing fittings, and stopcocks to gas analyzers and flow control systems. This book contains many generic references to them and Appendix 1 lists some useful sources. Anyone who has used a book of this sort knows that printed information on specific instrumentation rapidly becomes outdated, so to supplement and update the information in this book, my colleagues and I maintain a web site (www.respirometry.org) where additional information is available. The web site includes treatments of respirometry that for one reason or another are not included in this book, plus calculators, case histories, detailed descriptions, example citations, and links. Though by no means a substitute for this book, readers will find it a useful source of supplementary information. It includes a forum in which users of this book can allow expertise to diffuse along its concentration gradients.

Citations are always a contentious topic. I see this book as a concise guide to practical implementation of respirometry—a *vade mecum* as opposed to an encyclopedia. Thus, I have eschewed a plethora of citations, concentrating instead on papers that demonstrate relevant concepts or techniques clearly. These primary sources can be consulted for further citations. The book's web site also has a regularly updated list of citations that are ordered by chapters and topics in the book. These references illustrate use of the techniques described here in a wide variety of research contexts. If you want still more, search the Internet via, for example, Google Scholar. Depending on the perceived likelihood that many readers might use a description as a guide for setting up a system, I have included a greater or lesser level of detail in the text. Should you require more detail, you may wish to refer to the references I have provided or visit the companion web site.

This book is necessarily an idiosyncratic treatment of a wide and diverse field. If covered in the detail the field deserves, it would run to many volumes and be prohibitively cumbersome and expensive. I have left out much that in my judgment would be of limited utility, while at the same time repeating some information that I felt was important to present in context in different places. It is impossible, and probably undesirable, to write a textbook without offending anyone, or maybe everyone, by the sins of exclusion, inclusion, confusion, or effusion. That said, I welcome constructive comments about what to add or what to leave out in possible subsequent editions. I can be contacted via www.respirometry.org.

Finally, some acknowledgments are in order. I thank Robbin Turner for her support and friendship, which have made all the difference; my colleagues worldwide in the respirometry community, especially Mark Chappell and the late George Bartholomew; the students of my respirometry courses for their drive and enthusiasm; the Friday Harbor Laboratory and the Helen Riaboff Whitely Center for providing an ideal environment for the focused work that made this book possible; and Peter Prescott of Oxford University Press for encouragement and for keeping me, sometimes successfully, and against all odds, more or less on deadline.

CONTENTS

ABBREVIATIONS

AEE	Active energy expenditure. The EE of an animal engaged in activity of some kind, such as walking or running
ANCOVA	Analysis of covariance
BAT	Brown adipose tissue, a thermogenic organ in many mammals
BMR	Basal metabolic rate. The EE of an inactive, postabsorptive homeothermic endotherm such as a mammal or bird in its TNZ. Sometimes referred to as RMRt (resting metabolic rate at thermoneutrality)
BP	Barometric pressure
DEE	Daily energy expenditure. The mean EE over an entire circadian cycle
DIT	Diet-induced thermogenesis. A transient elevation of EE that occurs after ingesting food; also known as SDA
EE	Energy expenditure, the "real" metabolic rate of an organism. It can be calculated from VO_2 and VCO_2 (indirect calorimetry) or measured using direct calorimetry. Often expressed in kilocalories per hour or Watts (joules per second)
F	Fractional concentration of a given gas, such as Fo_2; usually distinguished by whether the concentration is measured entering a respirometer chamber (incurrent; F_i) or leaving it (excurrent; F_e)
FEP	Fluorinated ethylene propylene; an O_2-permeable polymer used for Clark probe electrodes
FQ	Food quotient; the ratio of CO_2 produced to O_2 consumed when burning a food in pure O_2
FR	Flow rate of air through a chamber or into a mask (mass flow rate, corrected to STP, is assumed unless stated otherwise)
HFM	Hollow fiber membrane; a CO_2- and O_2-permeable membrane with high surface area
HVAC	Heating, ventilation, and air conditioning system
IACUC	Institutional Animal Care and Use Committee; charged with maximizing the welfare (as they understand it) and minimizing the use of experimental animals in research
LCT	Lower critical temperature; the temperature below which the EE of a homeothermic endotherm will start to increase
MR	Metabolic rate; see EE
NIST	National Institute of Standards and Technology (USA)
PFA	Perfluoroalkoxy
po_2	Partial pressure of oxygen

PTFE   Polytetrafluoroethylene

PWM   Pulse width modulation; an electrical technique for adjusting power applied to a device by changing the width of constant-frequency pulses in its supply voltage

Q_{10} The ratiometric increase of an ectotherm's metabolic rate over a $10\,^\circ$C rise in body temperature; usually has a value close to 2.0

REE   Resting energy expenditure. The EE of a motionless animal, not necessarily post-absorptive or in its TNZ

RER   Respiratory exchange ratio. See RQ; mostly a pedantic synonym thereof, reflecting the possibility that an organism in metabolic or ionic disequilibrium may briefly show external VCO_2:VO_2 ratios outside the normal mitochondrial range

RH   Relative humidity, the percentage of water vapor pressure in the air relative to the maximum (saturated) water vapor pressure possible at the temperature of measurement

RMR   Resting metabolic rate; see REE

RQ   Respiratory quotient. The ratio of CO_2 emitted to O_2 consumed by the mitochondria; this typically falls within the range 0.7–1.0 for aerobic catabolism of fats and carbohydrates, respectively; proteins yield intermediate values based on their mix of amino acids

SDA   Specific dynamic action, a mistranslation of the already obscure term *spezifisch-dynamische Wirkung*; see DIT

STP   Standard temperature and pressure. Gas volume is typically corrected from ambient conditions to STP ($0\,^\circ$C and 101.325 kPa)

STPD   As STP, but with water vapor removed (dry)

TNZ   Thermal neutral zone; the temperature range over which an endothermic homeotherm such as a mammal or bird maintains its lowest EE

UCT   Upper critical temperature; the temperature above which the EE of a homeothermic endotherm will start to increase

V_{CO_2} Rate of CO_2 emission, often expressed as milliliters per minute, milliliters per hour, micromoles per second, and so on

V_{H_2O} Rate of water vapor loss, often expressed as milliliters per minute, as milliliters per hour, or in mass equivalents

V_{O_2} Rate of O_2 consumption, often expressed as milliliters per minute, milliliters per hour, micromoles per second, and so on

WVP   Water vapor pressure

1

Measuring the Fire of Life: A Brief History of Metabolic Measurement

Anyone who thinks that science follows a linear and orderly course from ignorance to enlightenment has not read much about the history of metabolic measurement. Tangled at its beginnings with alchemy, more than 100 years passed from its first inception before metabolic measurement began to resemble anything that makes sense to our modern worldview.

This is not the fault of metabolic measurement per se. Early thinking about air, fire, and life was marvelously unformed. It was known since prehistory that animal life depended on air, but what air was, and exactly *how* it supported life, was unknown. Leonardo da Vinci (1452–1519), in the *Codex Atlanticus*, stated that "Where flame cannot live, no animal that draws breath can live," but the evidence he considered, and the lines of reasoning he followed, are unknown. A rich tradition of mining and smelting, alloyed with alchemical lore, offered another perspective on air and its components. This is where things went wrong, though for the best of reasons. Charcoal, when burned, practically disappears, leaving behind an insubstantial smudge of ash. From this observation it was quite logically, and quite incorrectly, deduced that charcoal contained a vital substance that in and of itself supported fire. This substance, later called phlogiston by the proto-chemist Georg Stahl (1660–1734), had to be released into the air, and indeed, if the air developed a high enough concentration of phlogiston, it would no longer support combustion.

Within this paradigm it was easy to prove that phlogiston existed, even though it did not. One line of reasoning, mentioned above, was that its accumulation in air eventually made it impossible to add more phlogiston, making further combustion impossible. In another line of reasoning, a metal when strongly heated in air would "calcine," or turn into a powdery substance called the "calx" of that metal. Because fire purified things (everyone knew that), it was obvious that the original metal was really a mixture of the pure metal (which was the calx) and phlogiston, which was driven off by the heat. The calx was the "dephlogisticated" metal. This was proved easily by adding to the calx something that contains an abundance of phlogiston, such as charcoal (which you'll recall almost disappears when burned, and so has oodles of phlogiston) and heating the mixture strongly. Presto, the rephlogisticated metal reappears. Logically, the pure element

Measuring Metabolic Rates: A Manual for Scientists. Second Edition. John R. B. Lighton, Oxford University Press (2019).
© John R. B. Lighton (2019). DOI: 10.1093/oso/9780198830399.001.0001

(the calx) mixed with the phlogiston from the charcoal, reconstituting the metal. In other words, chemistry in those days was in the state of advancement that religion, politics, and philosophy were at then and remain today, with little hope for future improvement.

Not everything was in accord with the "facts," however. Before the general acceptance of the phlogiston theory lived Robert Boyle (1627–1691), a brilliant young Irishman among whose many contributions to science was the vacuum pump. He showed that candles, birds, and mice were all snuffed out if the pressure of the air around them was reduced to a low enough value. Thus, both combustion and life required something that was present in air. Enough empty space into which to dump phlogiston did not suffice. Boyle also showed that when metals were calcined, they did not lose weight after the phlogiston was driven off. Instead, they gained weight (which made sense in the phlogiston theory only if phlogiston had negative mass). His contemporary, the even more brilliant but lesser known John Mayow (1643–1679), asked the vital question that is obvious in hindsight: Does only one component in air support *both* life and combustion? Trapping mice under a bell jar in a saucer of water, Mayow watched as the water level rose. Plainly, the mice were consuming something in the air. They consistently died after the water rose by a certain amount. The same thing happened with a candle. Mayow had invented the first semi-quantitative respirometer. Placing both a candle and a mouse beneath the bell jar, first the candle would go out, and almost immediately afterward the mouse would die, suggesting that the same component supported both fire and life. He called the component of air on which both combustion and life depended "spiritus igneo-aereus" or sometimes "nitro-aereus." Amazingly, he hypothesized that this component enters the lungs and is passed into the blood, where the combination of nitro-aerial particles with combustible particles in the blood heats up the animal. This also happens, he hypothesized, in the animal's muscles during activity—including the activity of the heart, on which life depends. Then he became a Fellow of the Royal Society, married, and almost instantly died.

Joseph Priestley (1733–1804), who lived next door to a brewery, and perhaps not coincidentally was a good friend of Ben Franklin's, discovered that the gas produced by fermenting beer would extinguish flames. He also found that air that could no longer support combustion was composed partly of gas that would dissolve in water, identical to the beer-produced gas, which he called "fixed air." He found that fixed air could also be produced chemically by dripping oil of vitriol onto chalk, from which he made the first carbonated water. Priestley reasoned that the part of the air that was left over after supporting combustion and that could not dissolve in water as fixed air could, was the phlogiston. Nowadays we would call it nitrogen. He found that plants could eliminate fixed air, making it possible for the air to support life and combustion again. Continuing his research into gases, he found in 1775 that a different gas could be produced by heating red mercuric oxide under a burning-glass, and he was amazed to find that "a candle burned in it, and even better than in common air." Priestly also found that mice lived in this air for far longer than in normal air. And, fascinatingly, Priestley found that the air that remained after the mouse died dissolved in water and so was fixed air. Of course, we now know the gas that Priestley produced from mercuric oxide as oxygen. Influenced by current scientific theory, Priestley proceeded to totally misinterpret his observations. He proposed that the red mercury produced air, but air without phlogiston—dephlogisticated air. Red mercury became very popular among phlogiston researchers. One is tempted to ascribe much of the theory's appeal to the well-known effects of mercury, which inspired the guild of hatters.

It was only under the massive weight of observations made by the extraordinary Antoine Lavoisier (1743–1794) and the other "antiphlogistians" that the phlogiston theory finally buckled and crumbled, with Priestley remaining its last defender. The antiphlogistians realized that phlogiston was more parsimoniously explained as a gas, which Lavoisier named oxygen. The sophistication of the work performed by Lavoisier and his wife Marie-Anne Paulze is little appreciated. They measured oxygen consumption volumetrically in humans and animals, determined that oxygen consumption increased with animal size and with exercise, and discovered diet-induced thermogenesis. They invented indirect calorimetry, which measures metabolic rate (MR) via oxygen consumption. They also invented the first direct calorimeter, a device that used the body heat production (equivalent to energy expenditure (EE) or MR) of a guinea pig to melt ice. Knowing the volume of water melted from the ice per unit time, and the latent heat of fusion of ice (0.334 J g^{-1}), they could calculate, for the first time, the actual metabolic rate of an admittedly rather cold-stressed animal. This was a significant breakthrough and united the concepts of oxygen consumption and metabolic heat production. To give a less exotic example of this equivalence, a candle (also known in those days as a thurible, urceole, or veronica) burns lipids or waxes at the rate of about 6 g h^{-1}. A typical gram of candle wax yields about 45 kJ when combined with oxygen, so burning 6 g h^{-1} generates approximately 270 kJ h^{-1}, or about 75 W, of heat. This is similar to the metabolic rate of a typical human, giving a new facet of equivalence to Hamlet's "Out, out, brief candle." The range of Lavoisier's and Paulze's discoveries in this and many other fields was incredible, but their work was abruptly terminated during the French Revolution, when Antoine's flame was snuffed out after an administrator declared that "La Republique n'a pas besoin de savants"—the Republic has no need of geniuses.

It took another century for metabolic measurements to mature, chiefly via advances in manometry (Chapter 2). The next, massive advance in the field was the invention, especially during the middle of the twentieth century, of quantitative gas analyzers that were specifically sensitive to oxygen and carbon dioxide (Chapter 4). These allowed the metabolic rates of organisms to be quantified in real time, opening a new realm of applications in respirometry that are still expanding today.

Respirometry is an indispensable tool in many areas of science, and its direct human impact can be dramatic (think of the metabolic data required for understanding and treating the human obesity epidemic). In this respect, it is rather amusing to watch the sea change happening in the biomedical community, where whole-animal metabolic measurement used to be regarded as an intolerably incompetent relative, best kept out of sight and mind in the attic. Now the realization is dawning that molecular genetics and biochemistry, powerful as they are, offer a woefully incomplete understanding of the intact, functioning organism and that metabolic measurement offers a cornucopia of whole-animal insight that only a fool would ignore.

The great physiologist Max Kleiber (1961) memorably called aerobic metabolism "the fire of life." It is curious to sit back and think of the range of science encompassed by respirometry, the science of measuring the fire of life. I've just discussed the human history of respirometry, but the history of the atoms and molecules involved predates the formation of our planet. All of us have at some point engaged in the ancient tribal ritual, perhaps a half million years old, of squatting at a fire and taking sensual pleasure in its powerful all-senses mix of shimmering heat, evocative scent and taste, and hypnotic colors. But in the flickering darkness our modern mind sparks with thoughts our species has only just learned to think. We can think about who breathed out the carbon now being combined with oxygen in the fire, about the green life that cracks oxygen from water—the plants that make the fire possible, and about the sunlight that transforms the cold breath of the fire back into oxygen and gives rise to the β-glycoside skeleton of plants, repeating the cycle. Our thoughts turn to wider connections, such as red giant stars, their cool outer layers laced with carbon, oxygen, and nitrogen. At the core of the star, in a last race to the finish, carbon fuses with itself, oxygen, and helium to beget nitrogen, neon, magnesium, silicon, sulfur, and finally iron; and there the fusion process ends. With it dies the energy that maintained the vast sphere of the star against gravity, and falling inward, the star implodes into a black hole or neutron star and its outer layers rebound and supernova everything else—iron, oxygen, carbon, nitrogen, phosphorus, silicon, sulfur—out into the universe. Billions of years later, a speck of stardust squats before the fire and thinks of the greater fires that made its substance and that of the fire at which it warms itself. Through the heart of that speck pumps the iron that once blew apart the heart of a star, now meekly ferrying oxygen that the star created before its death. The fire and the speck marry oxygen to carbon—also a legacy of exploded stars—for the moment, and so create in one, heat alone and in the other, the fire of life.

Welcome to respirometry: measuring the fire of life.

2

Constant Volume and Constant Pressure Respirometry

This chapter deals with indirect O_2 consumption measurement techniques that rely on pressure changes to quantify the amount of O_2 consumed. These techniques, in one form or another, have been in wide use since the early twentieth century and are still being effectively used in some laboratories. The pressure changes, which derive from an organism's consumption of O_2 in a sealed container, are typically measured using a fluid manometer.

For these techniques to work, the CO_2 produced by the organism must be absorbed by a chemical scrubber such as potassium hydroxide (KOH) or Ascarite (see Appendix 1), and the temperature of the container (which has an alter ego as an extremely sensitive thermometer) must be kept constant, usually with a water bath. This combination of requirements limits the modern appeal of these techniques, but they are still capable of accurate results if used with care. As an incentive, suitable apparatus can often be found at little or no cost in department basements, in university surplus stores, or on Internet auction sites.

Nothing, of course, is perfect. The reader should be aware that manometric respirometry may not be as accurate as methods that directly measure O_2 and CO_2 concentrations (see Chapter 4). Interested readers should consult Van Voorhies et al. (2008) for the details.

There are two broad approaches to sealed-container respirometry that rely on pressure changes. In the older technique, the volume of the closed system is kept constant and the pressure change is quantified (the Warburg respirometer). It follows that the internal volume of the chamber and any attached plumbing must be precisely known and that the pressure changes must be accurately measured. These requirements can be quite difficult to achieve in practice. A slightly newer (but still comparatively ancient) technique requires knowledge of neither the chamber volume nor its actual pressure. The manometer is used only as an indicator of relative pressure, which is held constant by the operator, who adjusts the volume of the container by an amount equal to the volume of O_2 consumed, continuously returning the manometer's meniscus to its starting position. A typical manifestation of this technique is the Gilson respirometer, which is still widely used and will be described in detail first.

Measuring Metabolic Rates: A Manual for Scientists. Second Edition. John R. B. Lighton, Oxford University Press (2019).
© John R. B. Lighton (2019). DOI: 10.1093/oso/9780198830399.001.0001

THE GILSON CONSTANT PRESSURE
RESPIROMETER

For the latter half of the twentieth century the Gilson respirometer (Gilson, 1963) could be found in undergraduate laboratories around the world, giving students their first exposure to measuring metabolic rates. These instruments are not entirely extinct, and their relict survivors can still produce useful results. Unfortunately, not unlike other complicated hydraulic systems such as pipe organs, the Gilson respirometer demands some skill from its user. Its apparent complexity derives mostly from a massively parallel implementation of multiple elementary respirometers plus a controlled temperature water bath and agitator, all integrated into a large, intimidating, washing-machine-sized instrument. To raise the stakes, the potential for disaster lurks in multiple places, as we shall see.

Each measurement channel is quite simple. As a constant pressure device, use of the Gilson respirometer consists of the following steps:

1. Trap a specimen within a chamber.
2. Absorb any CO_2 produced by the specimen.
3. Measure the pressure change caused by the specimen's O_2 consumption.
4. Periodically adjust the chamber's volume to maintain constant pressure.

This all sounds simple enough. The chambers are typically small, conical glass flasks with a center well (Figure 2.1). For microbiological and enzymatic work, the chambers often have a side arm that allows a chemical, typically a respiratory substrate or a toxin, to be added to the mix if the flask is tipped. The center well is the home of the CO_2 absorbent, typically a 20 percent w/v KOH solution soaked into an accordion-folded square of filter paper that barely reaches the top of the well. To discourage the KOH solution from creeping and to prevent trapped air from leaking, the top of the well and all joining glass surfaces are coated with a thin layer of vacuum grease. This setup works well for immobile specimens, but mobile specimens will often interact with the grease or the KOH, making measurements problematic.

The pressure changes in the respirometer chambers are measured using manometers filled with dyed manometer oil. In operation, one side of each manometer connects to a respirometer chamber and the other connects to a thermobarometer, which is a chamber that controls for changes in temperature and barometric pressure. A single thermobarometer is used for all the respirometry chambers. On the side of the manometers attached to the respirometry chambers, small pistons driven by calibrated micrometer heads allow the volumes of each respirometer chamber circuit to be changed. Obviously, attaching or detaching a chamber will disturb its manometer, so a valve is provided that short-circuits the manometer when the valve is open. Another valve opens or closes the connection between the opposite side of the manometer and the thermobarometer. Opening or closing these valves in the wrong sequence can squirt manometer oil into

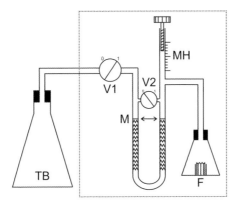

Figure 2.1 The Gilson constant pressure respirometer. A constant pressure difference is maintained between a thermobarometer (TB) and a flask (F) containing the experimental animal. To start the process, open main valve (V1) and close the manometer isolation valve (V2), thus connecting the manometer between TB and F. The manometer meniscus (M) is maintained at a constant level by manually adjusting the micrometer head (MH), calibrated in microliters, to displace the volume of O_2 consumed in the flask. A center well with fluted filter paper soaked in KOH solution absorbs CO_2 in the flask, so the change in pressure is caused by O_2 consumption alone. The section within the inside square is replicated multiple times.

the respirometer chamber, much to the detriment of the specimen. The manometer should only be included in the circuit when setup and equilibration are complete. The correct order of operation has usually been handwritten in large letters somewhere on the machine by a frustrated ex-operator, who has since quite possibly died of old age.

In use, after a long period of equilibration (typically an hour or so), the valve short-circuiting the manometer is carefully closed so that the manometer reads the pressure difference between the respirometer chamber and the thermobarometer. As the specimen uses (or produces) O_2, the change in pressure will pull the oil in the manometer in one direction or the other. The mission of the operator is to adjust the micrometer head so that the level of fluid in the manometer stays constant and to note the time and reading of each adjustment of the micrometer. Over a period of many minutes to several hours or days, a graph of micrometer head readings is built up, unless you backed up the micrometer below zero, in which case you broke its seal and the experiment is over. When an approximately constant slope is attained, the experiment can end well if the manometer is short-circuited before opening the valve that connects it to the thermobarometer; the experiment will end disastrously if that step is not followed.

The rate of change in the micrometer head's reading, which is generally calibrated in microliters, is equivalent to the rate of O_2 consumption (or production) by the specimen at the temperature and barometric pressure of the experiment. A simple adjustment yields standard temperature and pressure (STP)-corrected rates, so:

$$\text{Corrected } \mu L = (\mu L \times BP \times 273.15) / (101.325 \times T) \tag{2.1}$$

where BP is barometric pressure in kilopascals and T is the temperature of the specimen (= water bath) in Kelvin. After STP correction, the rate is usually expressed as
microliters per hour, micrograms per hour, or micromoles per hour. To convert from
microliters per hour at STP to micrograms per hour, multiply by 1.43 (32 g mol^{-1} divided
by 22.4 L mol^{-1} at STP). To convert from microliters per hour at STP to micromoles per
hour, multiply by 0.04464 (1/22.4 mol L^{-1} at STP).

The Gilson respirometer, though intimidating at first sight, is appealingly simple and
direct. Its intuitive directness is its primary advantage. You move a piston to displace the
oxygen used (or produced) by the specimen, giving you a direct physical measurement
of O$_2$ consumption. Moreover, there is no need to know the volume of the respirometer
chamber with any accuracy, unlike the case of the Warburg respirometer (discussed in
the section "The Warburg constant volume respirometer").

The single most serious disadvantage of the Gilson respirometer, apart from its potential for oily disasters, is the integrative nature of its readings. It shares this disadvantage
with all the other technologies described in this chapter. Fluctuations in MR are difficult
to distinguish from measurement noise. Of course, in the case of a microbiological or
enzyme preparation that is metabolically consistent, this is not a problem, but it can be
a massive drawback in the case of an animal that alters activity levels (Chapter 18) or
consumes O$_2$ or releases CO$_2$ intermittently, as do many tracheate arthropods (see Lighton,
1996, for a review). Activity increases MRs by large factors (by ~50 percent to > 800 percent,
depending on activity levels), often leading to substantial overestimates of nominally inactive MR. In a typical Gilson respirometer this problem is exacerbated because the water
bath ensures that the respirometer chambers are in a state of total internal reflection
with respect to the observer. The silvery chambers conceal their contents, making visual
activity checks difficult or impossible. A submersible mirror at the end of a wand can
help but is hardly user friendly. This is a problem because many arthropods, in particular,
are quite active when trapped in small spaces and may maintain elevated levels of activity
almost indefinitely. Brief periods of inactivity are difficult to perceive because of the
internal reflection issue and may be difficult to ascertain from the graph of consumed O$_2$
over time because of the variability of the data. Everyone who has used a Gilson respirometer will recall periods when their animal behaved like a plant or vice versa. Nowadays
you could probably publish that in *Science*; people used to be so skeptical.

Another disadvantage of the Gilson respirometer is the fact that the partial pressure of
oxygen (po_2) declines over the course of the experiment. This decline is usually insignificant
if the experiment is short, but long experiments or metabolically active specimens may
conspire to produce questionable data if the po_2 falls by more than about 1–2 percent, especially in the case of metabolic conformers, which show a simple dependence of MR on po_2.

Because the Gilson respirometer relies on changes in pressure within a closed vessel,
it is unsuitable for any specimen that may produce significant heat. This is seldom a
problem, but in the case of strongly endothermic animals such as bees and other flying
insects, wildly erratic readings should be expected as the insect warms up by shivering its
flight muscles or engages in futile cycling (see Heinrich, 1993) or cools down afterward.

The Gilson respirometer can be automated using a closed-loop control system. The meniscus level in the manometer can be sensed with a simple photoelectric sensor, and when the level rises above a threshold, a stepper motor can drive the micrometer head until the meniscus retreats to its proper place. If you know the number of pulses required to advance the micrometer head via the stepper motor to yield a given piston displacement, the rate of oxygen consumption can be directly recorded. There have been sporadic attempts in this direction (e.g., Johnson et al., 1982). However, the requirements for sensing the manometer meniscus and using a stepper motor to drive the micrometer head without inviting disaster have mostly resulted in idiosyncratic, one-of-a-kind set-ups. I don't recommend this approach for anyone trying to minimize their stress level. The Gilson company has moved into the pipetting and high-performance liquid chromatography (HPLC) businesses and does not mention its respirometers in its company history web page. Perhaps the memory is too traumatic.

THE WARBURG CONSTANT VOLUME RESPIROMETER

The Warburg constant volume respirometer was extensively used by Otto Warburg and his colleagues in Berlin, thus giving it its name, and it is still used in some laboratories. The device was first described by Joseph Barcroft and John Haldane (1902). It remains an accurate if tedious method for measuring the MR of small organisms, tissue samples, and microbiological preparations.

The Warburg respirometer consists of a sealed chamber, similar to the Gilson respirometer, complete with CO_2 absorbent in a well and a curved side arm. The chamber is connected to a manometer with one side open to the air (Figure 2.2). The manometer is made of precision-bore glass tubing, having an internal area as close as possible to 1 mm^2, for reasons outlined below (equation 2.2). The manometer is usually open at the bottom, where it connects to a short length of flexible tubing to which a clamp is applied. By adjusting a thumbscrew on the clamp, the level of fluid in the manometer can be changed to some convenient and memorable setting when a run is started. As O_2 is consumed, the level of the fluid in the manometer arm connected to the chamber rises (it might be objected that this changes the volume, but the effect is negligible and is accounted for in equation 2.2) while the other side falls. It's obvious from this description that (1) the temperature of the chamber must stay constant, which is usually achieved by immersing it in a water bath, and (2) the specific gravity of the manometer fluid is critical; the higher its density, the less the change in level for a given change in pressure. Aficionados of this technique endorse the use of Brodie's fluid (Brodie, 1910), which has a specific gravity such that a column 10 m high exerts a pressure of 101.3 kPa. Brodie's fluid can be prepared as follows: sodium chloride, 23 g; sodium tauroglycocholate, 5 g; water to 500 mL.

With a Brodie's fluid-filled manometer at hand, open the Warburg chamber to the outside air (such chambers are equipped with a tiny rotating valve on the side arm that

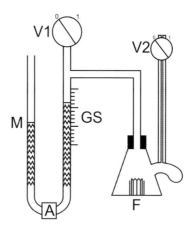

Figure 2.2 The Warburg constant volume respirometer. A manometer (M) senses the drop in pressure caused by O_2 consumption in the flask (F) containing an experimental animal. V1, Main valve, closed when reading; V2, flask isolation valve, closed when reading (shown in conceptual form; usually implemented as a glass tube rising to air level through a water bath containing the flask with ground-glass stopcock at the chamber arm); GS, graduated scale. The manometer fluid rises up the scale as O_2 is consumed in the flask. Fluted filter paper soaked in KOH solution is placed in a center well in the flask and acts as a CO_2 absorbent.

allows this; the valve has a long upward hollow stem so that it can be rotated while the chamber is immersed in a water bath). Adjust the clamp thumbscrew so that both arms of the manometer read 150 mm, which is the customary (though arbitrary) number. Seal the chamber and track the changes in the manometer readings over time. Barometric pressure changes will, of course, affect the readings but will generally not bias the data over the long run and over multiple runs.

The larger the chamber, the smaller the change in pressure for a given consumption of O_2. Thus, the volume of the chamber and its associated fitting must be exactly known, and such chambers usually have that value engraved on them, plus (often) the temperature at which that volume was determined. The equation relating manometer readings to O_2 consumption is as follows:

$$\mu L\ O_2 = h\left[\left(V_f\left[273.15/T\right] + V_q b\right)/P\right] \tag{2.2}$$

where h is the manometer reading in millimeters, V_f is the free volume of gas in milliliters within the chamber plus manometer up to the level of the fluid meniscus (the volume of the specimen is subtracted from the total volume), T is the chamber (= water bath) temperature in degrees Centigrade, V_q is the volume in milliliters of all liquids in the chamber, b is the solubility of the measured gas (usually O_2) in the chamber's liquids (if any) in microliters per milliliter at temperature T, and P is the reference barometric pressure in millimeters of manometric fluid, which, for Brodie's fluid, is 10,000 mm at nominal sea level. The rate of change in microliters O_2 is tracked over time, yielding the rate of O_2 consumption, as with the Gilson respirometer.

Variants of the Warburg respirometer with balancing thermobarometers and other ancillary instruments exist (e.g., Singh and Mathur, 1936), but the basic design has proved quite conservative. Its faults are broadly similar to those of the Gilson respirometer, with the added detraction that it is even less automatable. The reason for this is that a linear deflection of a fluid–air interface needs to be tracked and recorded. This is not easy to do accurately, even with modern technology, let alone when the device was in wide use.

THE DIFFERENTIAL PRESSURE RESPIROMETER

Reinventing the Warburg respirometer in differential form with modern differential pressure transducers can yield a low-tech but accurate means of measuring MR in small specimens. Here I describe the technique in enough detail to allow fabrication of a suitable device by anyone with moderate technical skills. My main purpose is to set the conceptual stage for coulometric respirometry, followed by stop-flow and flow-through respirometry. Nevertheless, practical implementation of the device would provide a low-budget but reasonably accurate means of sensitive MR measurement, suitable for educational use.

Consider a system comprising two sealed chambers of approximately equal size. An isometric, differential pressure sensor is mounted between them. For those unfamiliar with the terms, isometric means that pressure is measured without a significant displacement in the pressure transducer—which, being differential, responds only to the difference in pressures between the two chambers. In the absence of pressure changes between the two chambers, the sensor will read zero. Such sensors are available from many vendors (see Appendix 1). They have either analog or digital outputs and are easy to interface to data acquisition systems (Lighton, 1988). In principle, a manometer could replace the differential pressure sensor, yielding a differential-modified Warburg respirometer with no recording ability, but this modification would be pointless.

The pressure of air in the two chambers will initially be ambient barometric pressure, BP. Now imagine that all of the O_2 in one chamber is selectively withdrawn. The pressure in that chamber will decline to $BP(1 - Fo_2)$, where Fo_2 is the fractional concentration of oxygen in air (0.2094). At sea level, this corresponds to a differential pressure reading of about 21 kPa.

In the case of an organism consuming oxygen in one of the chambers, far smaller reductions in po_2 will be tolerated, and the organism is producing CO_2, which mitigates the pressure reduction effect and can build up to toxic concentrations. The latter drawback is addressed by using a CO_2 scrubber, as with the Gilson and Warburg respirometers (though see the section "Measuring respiratory quotient" for putting CO_2 buildup to constructive use). A reasonable maximum depletion of O_2 for most specimens might be

2 percent, corresponding to about 4 kPa maximum differential pressure, which is well within the range of inexpensive, off-the-shelf differential pressure sensors.

To generate quantitative data, the volume of the chamber holding the animal must be known, as should the volume of the animal. The latter can be approximated by dividing the animal's mass in grams by the density of the animal in grams per cubic centimeter. If you don't know the animal's density, 0.98 g cm^{-3} is a reasonable guess. Chamber volume can be calculated directly for chambers with simple geometries. The volume of more complex chambers can be measured by filling them with a fluid of known or easily measurable density, such as water or fine sand, and then weighing them empty and full and dividing the mass change by the density of the filler in grams per cubic centimeter. Tubing volume can be calculated from $(\text{tubing radius})^2 \times \pi \times$ tubing length.

Alternatively, you can use an attached syringe to alter the volume of the chamber and attached tubing by a known amount and measure the resulting pressure change. Suppose, for example, the entire volume of the chamber is withdrawn. This will halve the pressure within the chamber; if half the volume of the chamber is withdrawn, it will reduce the measured pressure by one-third. Thus $P_d = (BP \times V_s)/(V_s + V_c)$, where BP is the ambient barometric pressure, P_d is the differential pressure change, in the same units as BP caused by addition or withdrawal of sample V_s, and V_c is the chamber volume. Thus,

$$V_c = (BP \times V_s)/P_d \qquad (2.3)$$

In practice, BP is often measured with a barometer and P_d with a water manometer. It is worthwhile, at this point, to inject a note of caution about barometers. Most barometers designed for meteorological use are adjusted to read 101.3 kPa *at the altitude at which they are used, not at sea level.* Likewise, barometric pressure readings obtained from meteorological services are adjusted to normalize them to sea level. Such readings will not work for this calculation. You must explicitly obtain actual barometric pressure or convert the normalized pressure to actual pressure based on your altitude. Tables and formulas are available at www.respirometry.org.

Knowing the volume of the container (minus the volume of the specimen), the volume of O_2 in the container at the moment of sealing is $V_c(F_iO_2)$, while the pressure change corresponding to the removal of *all* O_2 is $BP(1 - F_iO_2)$. It follows that we can convert from a pressure change since the moment of sealing the chamber, Δp, to O_2 volume equivalents, thus:

$$\mu L\ O_2 = \left[V_c(\Delta p)(F_iO_2) \right]/\left[BP(1 - F_iO_2) \right] \qquad (2.4)$$

where V_c is the volume of the container (minus the volume of the specimen) in microliters and Δp and BP are in the same units. The resulting value can be converted to STP as for the Gilson respirometer, recalling that BP is barometric pressure at the time of chamber closure. This very simple treatment assumes absorption of CO_2 and neglects the effect of water vapor. By recording the total differential pressure change over time (for example, with any data acquisition system or even with a chart recorder), the rate of O_2

consumption is readily calculated as the slope of the pressure change over time when multiplied by the above expression.

The differential pressure technique suffers from all of the disadvantages of the techniques discussed thus far. It is, however, easily automatable, and if the chamber is suitably designed, activity or other relevant data can be recorded at the same time as pressure (see Chapter 18 for more on activity detection). Most important, exploring this technique allows us to ask an interesting question: What is to prevent us from keeping an experimental animal in the respirometer chamber indefinitely?

One way to achieve this goal would be to periodically open the thermobarometer to the outside air and open and flush the respirometer chamber before resealing it. This can, in principle, be achieved with one pneumatic solenoid valve for the thermobarometer and two valves for the animal chamber, one of which is connected to a low-flow-rate pump and the other to the outside air. When a specified pressure differential (e.g., 2 kPa) is sensed, the three solenoids can be opened, and a low-flow pump run for long enough to exchange the air within the respirometer chamber (Figure 2.3). This will typically take three time-constants for 95 percent exchange, where a time-constant in minutes is defined as the volume of the chamber in milliliters divided by the pump's flow rate in milliliters per minute (see also equation 8.1). The pump is then turned off and, after a pause, the solenoids are de-energized, after which the next round of measurements can begin. This is easy to achieve with any good data acquisition program such as LabView (see Chapter 20), but it has the disadvantage that the measurements and probably the animal are periodically disturbed.

This rather crude technique aside, the obvious factors preventing long-term measurements, neglecting longer-term issues such as food supply and waste management, are (1) declining po_2, (2) increasing pco_2 (unless you are scrubbing CO_2), and (3) water loss. Water loss can be mitigated by maintaining a high chamber humidity and CO_2 can be removed by scrubbing, but declining O_2 presents a bigger problem. This is explored further in Chapter 3.

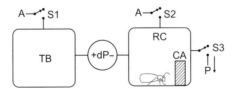

Figure 2.3 Differential pressure respirometer. The O_2 consumed by the experimental organism reduces the pressure in the respirometer chamber (RC; with CO_2 absorbent, CA) relative to the thermobarometer (TB). The pressure change is sensed by the differential pressure sensor (dP). A refined design with solenoid valves for replenishing the interior air of the respirometer chamber is shown. S1–S3, Solenoid valves, shown in the unenergized position. The solenoid valves are sealed unless a connection is shown. A, Connection to ambient air when the solenoid is energized; P, connected to pump P, pulling air through the respirometer chamber at approximately 200 mL min^{-1}, when energized. All solenoid valves may be energized simultaneously, together with the pump.

MEASURING RESPIRATORY QUOTIENT

If, when using any respirometry technique described in this chapter, you do not absorb CO_2, something interesting happens. If (and only if) the organism has a respiratory quotient (RQ; Vco_2/Vo_2) of unity (i.e., uses solely carbohydrates as fuel), no pressure change will take place. The organism will appear to have no MR at all. This gives a relatively quick and easy way to measure the RQ of an organism. To do so, measure the organism's apparent MR (MR_a) without CO_2 absorbent. Then measure its MR with CO_2 absorbent, which will always be larger. The RQ is then simply

$$RQ = (MR - MR_a)/MR \qquad (2.5)$$

This assumes, of course, that the organism's MR has remained constant. The expected range of RQ is 0.7–1 (pure lipid and pure carbohydrate catabolism, respectively; but see Chapter 10 for causes of atypical RQs). Likewise, MR measured in the absence of CO_2 absorbents gives a measure only of the *non-carbohydrate-fueled* metabolism of the specimen. This might provide interesting time-course data to demonstrate a dynamic shift in the catabolic allocation of respiratory substrates—for example, in animals entering a fasting or diapausing state. To the best of my knowledge, this measure has never been used explicitly in the literature.

CONTROLLING RELATIVE HUMIDITY

The requirement for a CO_2 scrubber for all techniques in this and Chapter 3 has an unexpected and potentially beneficial side-effect. If a liquid CO_2 scrubber such as a KOH solution is used, the water vapor pressure of the resulting solution can effectively control the relative humidity (RH) within the respirometry chamber. From data in Tartes et al. (1999), the following equation can be fitted:

$$KOH\% = 0.23262 + 1.2555(RH) - 0.63028(RH^2) + 0.013206(RH^3)$$
$$- 1.2750 \times 10^{-4}(RH^4) + 4.6201 \times 10^{-7}(RH^5) \qquad (2.6)$$

where KOH% is the percentage of KOH (w/v) in the KOH solution and RH is the desired percentage RH in the chamber. This equation is valid over the range 20–90 percent RH.

3

Coulometric Respirometry

Coulometric respirometry is an elegant but seldom used technique. It is based on constant volume and constant pressure in a sealed respirometer chamber and is remarkable on two grounds: its excellent accuracy and its potential for making long-term measurements on suitable organisms.

As discussed in Chapter 2, the maximum duration of a respirometric measurement using a closed respirometer chamber, assuming CO_2 absorption, is limited by the amount of O_2 available and the rate at which the enclosed organism consumes O_2. Coulometric respirometry sidesteps this issue by sensing the decline in pressure caused by consumption of O_2 within the chamber and generating O_2 in quantities sufficient to hold the pressure constant. For small animals, especially, electrolytic generation of O_2 works well, and it has the advantage that a known charge produces a precisely known molar quantity of O_2.

Coulometric respirometry was first described by N. T. Werthessen (1937), who used it to measure rat metabolism. Subsequent authors have used it chiefly for smaller organisms; excellent treatments can be found in Heusner et al. (1982) and Hoegh-Guldberg and Manahan (1995). Here I concentrate chiefly on small-animal use.

To generate O_2, a saturated solution of copper sulfate ($CuSO_4$) is electrolyzed using a platinum anode and a copper or platinum cathode. The electrochemical equation is

$$2CuSO_4 + 2H_2O \rightarrow 2H_2SO_4 + 2Cu + O_2 \tag{3.1}$$

Oxygen is generated at the anode, which should be sharpened to facilitate the production of tiny bubbles, while copper slowly plates onto the cathode. The volume of O_2 released is

$$nL\ O_2 = Q[V_m /(4F) \tag{3.2}$$

where Q is the charge of electricity discharged into the solution in coulombs, V_m is the molar volume of O_2 at STP (22.413×10^9 nL), and F is the Faraday constant (96,485 coulombs mol^{-1}).

You must therefore try to apply a precisely known charge into the electrolyte to generate a known quantity of O_2. Fortunately, this is quite easy to do. The charge stored by a capacitor is precisely CV, where C is capacitance in farads and V is the voltage stored by the capacitor, assuming that it is discharged to 0 V. This is not advisable in the case of

Measuring Metabolic Rates: A Manual for Scientists. Second Edition. John R. B. Lighton, Oxford University Press (2019).
© John R. B. Lighton (2019). DOI: 10.1093/oso/9780198830399.001.0001

electrolysis, because $CuSO_4$ will not decompose electrolytically below an applied potential of 2 V. Thus, you must charge a capacitor to some initial voltage, V_i, which is significantly greater than 2 V, stop the discharge at V_s, which is slightly greater than 2 V, and multiply by the coulometric constant (58,073) to obtain the volume of O_2 released per partial capacitor discharge:

$$nL\ O_2 = (58,073C)\left(V_i - V_s\right)$$

(3.3)

High-quality capacitors with precisely known values are easy to obtain (see Appendix 1) and they remain highly stable over time. The steps are as follows:

- Measure the pressure differential between the respirometer chamber and an adjacent thermobarometer.
- When the pressure in the respirometer falls below some set point, discharge a capacitor from V_i to V_s through a pair of platinum electrodes in a chamber, adjacent to and connected to the respirometer chamber, containing $CuSO_4$ electrolyte.
- Repeatedly discharge the capacitor and count the number of discharges until the pressure in the respirometry chamber reaches the set point again.
- It is then easy to compute the volume of O_2 added to the chamber, using the above equations.

Plainly, the system has two critical components, and these are the means of sensing pressure differentials and of discharging the capacitor accurately and repeatably into the electrolyte.

Unlike the differential pressure sensor described in Chapter 2, the pressure sensor used for coulometric respirometry does not need to be either linear or accurate because it is simply a threshold sensor that operates in a closed-loop control system. It does, of course, need to be stable. Under these loose selective constraints, several different pressure sensors have evolved.

Perhaps the most complex (but also the most precise) is that described by Heusner et al. (1982) and later elaborated by Hoegh-Guldberg and Manahan (1995). Basically, a column of saturated $CuSO_4$ in water in a glass tube senses the differential pressure between the respirometer chamber and the thermobarometer. Within the glass tube a Teflon ring constrains the meniscus on the respirometer chamber side, and pointing down into the meniscus is an electrolytically sharpened platinum electrode. Ingeniously, this electrode is also the anode of the electrolysis system. When the electrode contacts the $CuSO_4$ solution, a circuit is completed, triggering a second circuit to discharge a capacitor through the electrode into the solution, releasing O_2. A succession of discharges occurs until the accumulated O_2 drives the meniscus back and breaks the contact (Figure 3.1). It's important to note that the roles of the platinum electrode as a contact sensor and as an anode are separate; the two operations do not occur simultaneously. The meniscus–Teflon interface is one of exquisite sensitivity, as demonstrated via trigonometric analysis by Heusner et al. (1982). Suitable circuitry for implementing the

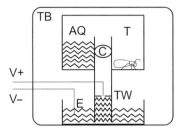

Figure 3.1 A coulometric respirometer conceptually similar to those described by Heusner et al. (1982) and Hoegh-Guldberg and Manahan (1995). The respirometer is enclosed in a thermobarometer (TB), which is normally placed in a water bath. The O_2 consumption of either aquatic (AQ) or terrestrial (T) organisms can be measured. CO_2 produced by the organisms is absorbed by a scrubber (C). As O_2 is consumed, the electrolyte (E) rises up to the Teflon washer (TW) and touches the anode (V+), completing the electrical circuit and allowing electrolysis to take place, producing O_2 in pulses created by repeatedly discharging a capacitor until the electrolyte meniscus is driven out of contact with the anode. The number of discrete pulses over time, each corresponding to the production of a known and fixed amount of O_2, allows O_2 consumption rate to be calculated.

capacitor discharge technique is described by Heusner et al. (1982); Hoegh-Guldberg and Manahan (1995) describe the full implementation of such a system, including computer-assisted data acquisition and analysis.

A number of potential sources of error exist in coulometric respirometry, though their predicted magnitude is small (Heusner et al., 1982). Hoegh-Guldberg and Manahan (1995) looked with great thoroughness at the actual magnitude of the errors in practice, expressed as deviation in actual O_2 production from theoretically expected values. To assess actual O_2 production, they used both volumetric and actual analytical O_2 measurements, and in each case, they assessed production rate using least-squares regression analysis with the number of discharges on the X axis. The theoretical volume of O_2 per discharge in their system was 4.20 pmol. Using volumetric techniques, they measured a value of 4.19 pmol per discharge, and using direct O_2 analysis, they measured a value of 4.18 pmol per discharge. In the hands of careful workers, this technique is plainly capable of outstanding accuracy.

A simpler yet in some ways more versatile approach is described in detail by Tartes et al. (1999), in a design adapted from Sláma (1988). Here, the pressure sensing and the electrolysis are separated. A thin capillary of ethanol senses pressure, and its position is determined photoelectrically by a light-emitting diode (LED)/phototransistor gate (Figure 3.2).

Photoelectric detection of meniscus position is capable of great precision, yet it is mechanically simple to set up and noninvasive. Sláma (1988) used a sensitive differential pressure transducer instead of an ethanol meniscus, but the key principle remains equivalent; either technique is intrinsically analog rather than digital (on/off), and this offers an important advantage over the electrical contact technique. The excursion of the meniscus through the light gate (or the deflection of the diaphragm of the differential pressure transducer) gives rise to a range of electrical values that can be used to determine

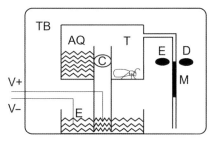

Figure 3.2 An alternative, continuously recording coulometric respirometer design, extensively simplified from Tartes et al. (1999). A thin capillary manometer (M) interrupts a light beam between a photoemitter (E) and a photodetector (D) as the experimental organism, whether aquatic (AQ) or terrestrial (T), consumes O_2. A CO_2 scrubber (C) absorbs CO_2 from the experimental organism and maintains a set relative humidity. A graded current, proportional to manometer displacement, flows via V+ and V− into a sharpened platinum anode (a copper cathode is used). The current passing through the electrolyte (E) is proportional to the displacement of the ethanol meniscus relative to the light gate; the greater the displacement, the greater the current, and the greater the counteracting production of O_2. The current passing through the electrolysis system is recorded and can be calibrated in O_2 consumption rate, allowing continuous records to be kept. A thermobarometer (TB) reduces sensitivity to barometric pressure and temperature changes.

how far the pressure is away from a given set point. Thus, because the pressure can now be sensed over a range of values (as opposed to in a binary, on/off fashion as with Heusner et al., 1982), an important new development is possible. The further the chamber pressure drops below its set point, the more current that can be passed through the electrolytic solution in a closed-loop control system, thus generating proportionately more oxygen to push the chamber pressure back to its set point. By recording the current delivered to the electrolysis system, this arrangement essentially gives a continuous readout of O_2 consumption rate. All that is needed is the conversion factor to convert from current to STP-corrected O_2 volume. This conversion factor is 209.5 μL O_2 mA^{-1} h^{-1} (Taylor, 1977). Using this system, Tartes et al. (1999) were able to measure O_2 consumption rates down to about 0.2 μL h^{-1} on a continuous basis, provided the anode was sharpened optimally so that the requisite tiny bubbles could be generated.

At present, there is no work of which I am aware involving coulometric respirometry of larger animals such as vertebrates. It can be argued that there are practical and logistical reasons for this, though large-scale coulometric production of O_2 is hardly infeasible; for instance, it is routinely used in crewed spacecraft. The real reason lies in the fact that metabolic data can be far more easily obtained in other ways, in the case of large animals at least, and I explore these approaches later, especially in Chapters 10 and 11.

That said, coulometric respirometry is by far the most sensitive (at least semi-) mainstream technique for measuring O_2 consumption in small animals, particularly over long intervals. It deserves to be more widely used. As Heusner et al. (1982) point out,

"The present use of coulometric methods for metabolic studies does not do justice to their considerable advantages over the conventional gasometric methods. In particular, they are ideally suited for automatic long-term recording of instantaneous O_2 consumption in very small animals. Their calibration is very stable in time and independent of the geometry of the respiratory chamber. The respirometer can be autoclaved, and the electrolytic O_2 is sterile; therefore, the recording of O_2 consumption in tissue or organ cultures is possible under aseptic conditions and over extended periods of time." (p. 185)

Humans are conservative beasts and, being human, so are scientists. The primary factor preventing the wider use of coulometric respirometry is simply that commercial setups are not yet available. Perhaps this will change in the future and this promising technique will become more accessible and popular.

4

Constant Volume Techniques
Using Gas Analysis

Chapters 2 and 3 described a variety of constant volume and constant pressure techniques that rely on pressure changes (or keeping pressure constant) to provide an indirect index of O_2 consumption. All these techniques suffer from the drawback that information about CO_2 emission is either lost or, at best, must be obtained by serial runs with and without a CO_2 absorbent. They also require stable temperatures and are finicky to work with.

More recently (from the latter half of the twentieth century onward), the growing availability of gas analyzers that can directly detect and quantify levels of O_2, CO_2, and other gases has led to a revolution in constant-volume gas analysis. Technical information on these analyzers can be found in Chapter 16. These analyzers allow the accurate and, if desired, automated determination of O_2 consumption and/or CO_2 production rates from a wide variety of samples, ranging from small vertebrates to arthropods (including the important model organism *Drosophila melanogaster*) and still smaller invertebrates (including another important model organism, *Caenorhabditis elegans*) to microbes. Depending on the implementation, the enclosure systems can vary in complexity from a syringe with a three-way valve to dozens of respirometer chambers switched under computer control. All implementations require one or more gas analyzers, usually with a means of providing accurately metered flow rates (see Chapter 17 for more information on flow measurement).

As a general rule, this chapter is relevant to anyone who wants to measure MRs of small organisms from bacteria up to about the size of a frog or a medium-sized lizard. Larger organisms, and those with higher MRs such as mammals or birds, are usually measured using flow-through respirometry. This is mostly because these organisms use O_2 and release CO_2 at rates that are too fast for chambers of a practical size to be sealed for a practical period of time.

I'll start by describing an elegant technique, static injection analysis, requiring only an O_2 analyzer, that is capable of measuring not only O_2 consumption rate but also, with a couple of small additional steps, CO_2 production rate and H_2O loss rate. Developed over several years in George Bartholomew's laboratory at the University of California, Los Angeles (UCLA), it is well summarized by David Vleck (1987). Following the arguments

Measuring Metabolic Rates: A Manual for Scientists. Second Edition. John R. B. Lighton, Oxford University Press (2019).
© John R. B. Lighton (2019). DOI: 10.1093/oso/9780198830399.001.0001

in this example will help you understand later examples, including those in later chapters involving flow-through gas analysis. Perhaps the most important point is that the oxygen content of a chamber, being depleted by an organism by a small amount from a huge initial concentration of close to 21 percent, is exquisitely sensitive to dilution—not only by water vapor but also by the CO_2 produced by the organism.

After dealing in detail with static injection analysis, I examine dynamic injection analysis, which is a variant of the former technique that allows the easy addition of CO_2 analysis. Next, I examine manual and automatic bolus integration techniques, which require more data acquisition and analysis power but have a number of significant advantages. Again, these techniques will introduce you to many of the concepts that are developed in later chapters on flow-through respirometry.

STATIC INJECTION ANALYSIS

Consider a closed chamber with a known volume, V. The volume excludes the volume of an organism in the chamber whose metabolic rate you want to measure (see Chapter 2 for chamber volume and organism volume determination methods). Within that chamber, the fractional concentration of O_2 at the start of the experiment is

$$F_iO_2 = V_iO_2 / V \qquad (4.1)$$

where V_iO_2 is the volume of O_2 initially present at the start of the experiment before the organism within the chamber has consumed any O_2. At the end of the experiment, F_iO_2 has declined to F_eO_2, defined thus:

$$F_eO_2 = (V_iO_2 - VolO_2)/(V - VolO_2 + VolCO_2 + VolH_2O) \qquad (4.2)$$

where $VolO_2$ is the volume of O_2 consumed by the creature, and $VolCO_2$ and $VolH_2O$ are the volumes of CO_2 produced and water vapor lost by the organism. These values are all determined by gas analysis—in this case, using only an O_2 analyzer.

The chamber can be of almost any size and design. Typically, the practical size range is from a few liters to 5 mL or so. It must have a sampling port and, of course, must seal well, which can be tested by applying a small positive or negative pressure to the sampling port using a syringe and monitoring the pressure change in the container using a water manometer or a low-pressure gauge (see Appendix 1). After the initial disturbance, the pressure should not change significantly if the seal is adequate and the chamber temperature remains constant. The chamber can be as fancy as you wish. Sampling ports are easily made from a bulkhead-mount Luer taper female fitting and a three-way Luer stopcock (see Appendix 1).

Obviously, a sample of air needs to be withdrawn from the chamber at the end of the experiment. Depending on the design of the chamber, this can be accomplished by withdrawing a subsample into a syringe or, if the chamber *is* a syringe (very practical for

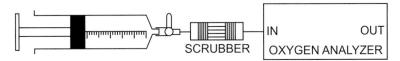

Figure 4.1 Static injection technique for constant volume respirometry. A syringe fitted with a three-way valve is slowly emptied through a small scrubber column (typically Drierite® and Ascarite) into an O_2 analyzer and a static reading is taken of the resulting O_2 concentration. The syringe can contain the organism being measured or contain a subsample from a larger chamber. The scrubber can be made from a short length of tubing. Its internal volume should be as small as possible relative to the injected sample volume.

small organisms), by squeezing a sample directly from it. In the former case, the sample should be slightly compressed after withdrawing it and sealing the sampling syringe to prevent the entry of ambient air. The subsample syringe is normally fitted with a three-way Luer stopcock, which allows the syringe to be sealed completely after the sample is withdrawn but before the connection to the sampling port is broken. Figure 4.1 illustrates the principle.

For a zirconia-cell or paramagnetic O_2 analyzer (see Chapter 16 for O_2 analyzer types), the sample can then be slowly injected into the analyzer and a reading taken. Fuel-cell O_2 analyzers are slightly trickier to use in this application because of their slower response times and because they consume O_2, which causes static readings to decline slowly over time as a boundary layer of O_2-depleted air develops at the sensor's membrane surface. Fuel-cell O_2 analyzers can still be used for static samples if you are aware of these effects, but they are more suited to injection of samples into a flowing air stream, which is a more versatile technique that is covered in the section "Option A." Whatever analyzer is used, it should be capable of resolution to at least 0.01 percent and preferably to 0.001 percent. Note that this technique is not recommended for use with a CO_2 analyzer as well as an O_2 analyzer because most CO_2 analyzers have significant internal volume, making static injections prone to mixing errors, especially if placed in line with another analyzer (this is less of a problem with the dynamic injection technique and not a problem at all with the bolus integration technique, which are described in sections "Dynamic injection analysis" and "Automatic bolus integration," respectively).

Let us analyze what we expect from our O_2 analyzer. From equation 4.2, the volume of O_2 consumed by the organism is

$$\text{Vol}_{O_2} = [V(F_iO_2 - F_eO_2) - F_eO_2(\text{Vol}_{CO_2} + \text{Vol}_{H_2O})]/(1 - F_eO_2) \qquad (4.3)$$

To obtain the rate of O_2 consumption, simply divide the volume of O_2 consumed (after STP correction; see equation 2.1) by the duration of the organism's confinement before taking the sample. From this, it follows that you need to know Vol_{CO_2} and Vol_{H_2O} before you can calculate Vol_{O_2}. There are three distinct strategies for coping with this problem, assuming, as we do for the moment, that only an O_2 analyzer is available. Either (1) simply inject the sample into the analyzer and make certain assumptions (which we'll come to in a moment), (2) absorb water vapor, or (3) absorb both water vapor and CO_2. (Absorbing only CO_2 is not practical because CO_2 absorbents interact with water vapor.) In the last two cases, the measured fractional concentration of O_2 must increase because the diluent gases, H_2O or both H_2O and CO_2, are removed. Following the conventions of Vleck (1987), I call the fractional concentration of O_2 after removal of water vapor $F'o_2$, and the fractional concentration of O_2 after removal of both water vapor and CO_2 $F''o_2$. I follow this convention throughout this book. What are the consequences of these different scrubbing regimes?

Option A: no scrubbing of carbon dioxide or water

First examining the case where no scrubbing of the sample gas takes place, there is no direct way of ascertaining $Volco_2$, but we can assume a value based on $Volo_2$ because almost all animals produce CO_2 at a ratio somewhere between $0.7(Volo_2)$ and $Volo_2$ (where the coefficient is the RQ, by definition). Thus, we can substitute $RQ(Volo_2)$ for $Volco_2$ in equation 4.3:

$$Volo_2 = [V(F_i o_2 - F_e o_2) - F_e o_2 [VolH_2O]]/[1 - F_e o_2 (1 - RQ)] \qquad (4.4)$$

It can be shown that the maximum error introduced by this assumption is about 3 percent if we assume an RQ of 0.85 (Vleck, 1987). However, because of its diluting effect on $F_i o_2$, which, without knowing its magnitude, cannot be distinguished from $Volo_2$, $VolH_2O$ is a far more serious source of error if its value is not known. Thus, if not considered, $VolH_2O$ will always overestimate $Volo_2$ by a factor of

$$V(F_i o_2 - F_e o_2)/[V(F_i o_2 - F_e o_2) - F_e o_2 (VolH_2O)] \qquad (4.5)$$

As an example, if we assume that the sample air is saturated with water vapor at $35°C$, the $VolH_2O$ at sea level will be the chamber volume multiplied by saturated vapor pressure of water in kiloPascals divided by 101.3 kPa, or $0.055(V)$. If $Volo_2$ is $0.01V$ (a reasonable value), $F_i o_2$ is 0.209, and RQ is 1, the overestimate is a factor of 2.05. Thus, you need to know $VolH_2O$, and in the absence of water vapor analysis, you need to either fix $VolH_2O$ at zero by including a desiccant such as Drierite in the chamber or fix it at saturation by including enough free water to saturate the air in the chamber. In the latter case you can calculate $VolH_2O$ as shown above, using either a table of saturated water vapor pressure versus temperature or a water vapor pressure (WVP) calculator (available at www.respirometry.org).

Option B: scrubbing only water

Scrubbing H_2O from a gas sample is quite easy; for small samples ($\sim$ 10–60 mL), an adequate scrubber can be made from a 3-mL syringe with a small cork on one end through which the incurrent gas stream passes, or even a section of tubing, filled with Drierite (anhydrous $CaSO_4$) or silica gel. Under these circumstances,

$$F'_iO_2 = V_iO_2/(V - V_iH_2O) \tag{4.6}$$

and

$$F'_eO_2 = (V_iO_2 - VolO_2)/(V - V_iH_2O - VolO_2 + VolcO_2) \tag{4.7}$$

Solving for $VolO_2$,

$$VolO_2 = [(V - V_iH_2O)(F'_iO_2 - F'_eO_2) - F'_eO_2(VolcO_2)]/(1 - F'_eO_2) \tag{4.8}$$

Or, if we substitute $(RQ)(VolO_2)$ for $VolcO_2$,

$$VolO_2 = [(V - V_iH_2O)(F'_iO_2 - F'_eO_2)]/[1 - F'_eO_2(1 - RQ)] \tag{4.9}$$

Now the only problem is the initial volume of H_2O, which can either be measured or be set to zero by scrubbing water vapor from the air that initially filled the chamber. Alternatively, you can calculate this volume, provided that initial water vapor pressure is known, as shown above. Water vapor analyzers with direct readout in kilopascals of water vapor pressure do exist (see Chapter 16) or, alternatively, relative humidity can be measured, divided by 100, and multiplied by the saturated water vapor pressure at the temperature of measurement:

$$WVP = (RH/100)\,SWVP \tag{4.10}$$

where WVP is water vapor pressure, SWVP is saturated water vapor pressure at the temperature of the RH measurement (see equation 16.7), and RH is relative humidity in percent. Ignoring F_iH_2O (i.e., assuming that it is zero when it isn't) can cause errors of up to 5 percent, depending on water vapor pressure and barometric pressure, and is not recommended.

You may already be asking, "What if we split an initial air sample and analyze half with a water scrubber and half without? Wouldn't the difference between measured values of F_iO_2 and F'_iO_2 give us a way of measuring V_iH_2O?" Yes. In fact,

$$V_iH_2O = V(F'_iO_2 - F_iO_2)/F'_iO_2 \tag{4.11}$$

This gives you a way to measure water vapor pressure without using a water vapor analyzer, provided you are willing and able to split the sample. Note that we are making the important assumption that F_iO_2 is stable. If you have reason to doubt that it is not (e.g., if it is measured in a populated room), then use a carboy to damp out concentration fluctuations (Figure 4.2).

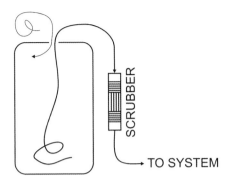

Figure 4.2 Reducing fluctuations of F_iO_2 by using a carboy or other large container. The carboy volume can be 10–50 L or more. A scrubber column can be used to remove water vapor and CO_2 from the sampled air stream if desired. Best results are obtained if the carboy is situated outside the building or supplied with outside air from an external source such as an aquarium pump located outside a window.

Option C: scrubbing both water and carbon dioxide

When we scrub CO_2 in addition to H_2O, we are once more removing a diluent gas, which increases measured F''_eO_2. Now,

$$F''_iO_2 = V_iO_2/(V - V_iH_2O - V_iCO_2) \tag{4.12}$$

and

$$F''_eO_2 = (V_iO_2 - VoI_{O_2})/(V - V_iH_2O - VoI_{O_2} - ViCO_2) \tag{4.13}$$

Solving for Vol_{O_2},

$$Vol_{O_2} = [(V - V_iH_2O - V_iCO_2)(F''_iO_2 - F''_eO_2)]/(1 - F''_eO_2) \tag{4.14}$$

Note something important: Vol_{O_2} in this equation is independent of $VolCO_2$. It is still, unfortunately, necessary to know V_iH_2O and V_iCO_2; however, the latter term is usually too small to matter. V_iCO_2 can nevertheless be measured by its dilution effect on F_iO_2, analogously to the technique used above to measure F_iH_2O:

$$V_iCO_2 = V(F'_iO_2)/[(1/F'_iO_2) - (1/F''_iO_2)] \tag{4.15}$$

Of course, this requires splitting the gas sample into three parts, and injecting one with no scrubber, one with an H_2O scrubber, and a third with both an H_2O and a CO_2 scrubber. This is not a great hardship and does not compromise accuracy provided that the volumes of the scrubber columns are small (e.g., < 20 percent) of the split volume of the sample.

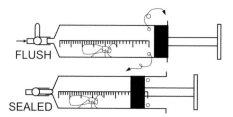

Figure 4.3 Modifying a syringe to allow pre-enclosure air flow for establishing a stable, repeatable F_iO_2. Small holes are drilled into or melted from the syringe barrel just within the outermost limit of the fully withdrawn plunger. Air is pumped into the syringe and escapes through the holes. To seal the syringe, disconnect the air source, press the plunger in to a known graduation mark so that the holes are isolated from the sample, and seal the sample with the three-way valve.

It is possible to measure not only O_2 consumption but water vapor loss and CO_2 production using the three-way split-sample technique. Experts agree that the three-way method has much to commend it.

$$\text{Vol}_{CO_2} = [(F''_eO_2 - F'_eO_2)(V - V_iH_2O - \text{Vol}_{O_2}) - F''_eO_2(V_iCO_2)]/F'_eO_2 \qquad (4.16)$$

Now, knowing Vol_{CO_2}, we can calculate V_{H_2O}:

$$\text{Vol}_{H_2O} = [(F'_eO_2 - F_eO_2)(V - \text{Vol}_{O_2} + \text{Vol}_{CO_2}) - (F'_eO_2(V_{H_2O}))]/F_eO_2 \qquad (4.17)$$

By far the easiest setup for using this technique, especially for multiple samples, is to arrange for a source of dry, CO_2-free air, for example, by attaching a Drierite/Ascarite/Drierite column to the output of a small diaphragm pump, and then using that scrubbed air stream to flush each chamber before starting measurements. This eliminates the requirement to know F_iH_2O and F_iCO_2. The flow rate of the scrubbed air stream should be high enough that the steady-state effect of the subject animal is negligible. If (as is often the case with this technique) insects or other small organisms are the study animals, they can be directly housed in syringes with stopcocks, and with a simple modification (Figure 4.3), the syringe can be flushed with minimal disturbance of the animal.

DYNAMIC INJECTION ANALYSIS

This variant of static injection analysis makes use of an air stream that flows continuously through the gas analyzer(s); it is described by Bartholomew et al. (1985). A diagram of such a system is shown in Figure 4.4. Samples are injected into an air stream that is assumed to be equivalent in O_2 concentration (and CO_2 and H_2O vapor concentration, if these are not scrubbed) to the air in the chamber(s) at the start of the enclosure period. This is usually a safe assumption if air from outside the building is used or if a source of compressed or bottled air is used. It is usually best to scrub CO_2 and water vapor from

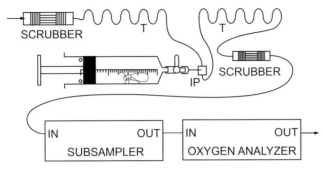

Figure 4.4 Dynamic injection technique for constant volume respirometry. Air, scrubbed of water vapor and CO_2, is continuously drawn through an O_2 (or CO_2) analyzer at a flow rate of 50–100 mL min^{-1} by a subsampling pump. The exact flow rate does not need to be known. A sample from a syringe (either containing an animal or containing a sample from a larger chamber) is injected into the air stream via the injection port (IP). Note the lengths of tubing (T), which ensure that the entire injected volume is drawn through the analyzers in a smooth bolus. The injected air is scrubbed by a second, minimum-volume scrubber.

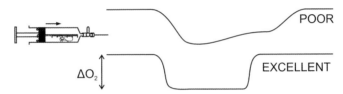

Figure 4.5 Oxygen depletion profiles for a dynamic injection setup. It is best to aim for a combination of injection rate, injection volume, and background flow rate that yields a stable, flat depletion profile, as shown in the second profile. The first profile shows poor displacement of the air flowing through the system with the sample air injected from the syringe.

the air entering the chambers. If this is done, the equations are simplified (see equation 4.18) and those two major sources of variation are eliminated.

The flow rate through the analyzers does not have to be exactly known and is adjusted so that injecting a sample of O_2-depleted air into the tubing through an injection port creates a smooth, flat-bottomed concentration profile (Figure 4.5) that is swept through the analyzers in turn (50–100 mL min^{-1} is typical). A graphing data acquisition package is invaluable here. Some experimentation with injection volume and injection rate will be required. Make sure that there is enough volume in the tubing system to allow the requisite volume of sample to be injected (when using this technique, I have ensured this by adding coils of 6 mm internal diameter (i.d.) tubing to either side of the injection port, with a calculated internal volume exceeding the largest anticipated injection volume). Before entering the analyzers, the air stream is scrubbed of water vapor and/or CO_2, depending on the analyzers being used and the aims of the experiment.

As with static injection, the volume of the chamber must be accurately known, and the volume of the organism should be subtracted from it. Equations 4.18 and 4.19 return results in volumes of O_2 consumed or CO_2 produced; divide these volumes by enclosure times to yield rates after STP correction (equation 2.1).

If H_2O and CO_2 were scrubbed from the air entering the chambers and are scrubbed again before O_2 analysis, RQ-independent $Volo_2$ can be calculated:

$$Volo_2 = V(F''_io_2 - F''_eo_2)/(1 - F''_eo_2) \tag{4.18}$$

If CO_2 is not scrubbed before analysis and is not measured separately (but was scrubbed, together with water vapor, from the air entering the chambers), then

$$Volo_2 = V(F'_io_2 - F'_eo_2)/[1 - F'_eo_2(1 - RQ)] \tag{4.19}$$

Again, assuming RQ = 0.85 ensures that $Volo_2$ will be accurate to about 3 percent in the worst case.

If CO_2 is measured separately (but was scrubbed, together with water vapor, from the air entering the chambers), then two ways of proceeding are open to you:

(1) Pass the sample through the CO_2 analyzer first, then scrub CO_2 and measure O_2, in which case $Volo_2$ is as described above, and $Volco_2$, which is very slightly elevated because of the removal of O_2 from the air stream by the animal, is

$$Volco_2 = [V(F'_eco_2) - F'_eco_2(Volo_2)]/(1 - F'_eco_2) \tag{4.20}$$

However, it is obvious that F'_eco_2 is usually very close to zero and that $Volo_2$ is generally very small compared to V. If F'_eco_2 is kept within moderate limits (try substituting values and see for yourself), equation 4.20 simplifies to

$$VolCO_2 = V(F'_eco_2) \tag{4.21}$$

(2) With the above simplification in mind, you can optionally remove the CO_2 scrubber before O_2 analysis. First calculate Vco_2 as above, and then use that value to dilution-correct Vo_2:

$$Vo_2 = V[(F'_iO_2 - F'_eO_2) - F'_eO_2(F'_eCO_2)]/(1 - F'_eO_2) \tag{4.22}$$

As you can see, it is generally easier to scrub CO_2 from the airstream before O_2 analysis. If you choose not to scrub CO_2 from the air entering the chambers, then F'_eco_2 in equations 4.21 and 4.22 should be replaced with $(F'_eco_2 - F'_ico_2)$.

Manual bolus integration

Manual bolus integration is superficially similar to dynamic injection analysis, with two critical differences. First, the injected volume must be precisely known, and second, the flow rate of the background air (or gas) through the system must also be known. In addition, computerized data acquisition and analysis are required. These are minor obstacles compared to the benefits derived from this technique and especially its automated sibling, discussed below. In the case of manual bolus integration, the amount and the speed of sample injection are not critical (provided the volume of the sample and chamber are known), and even very small volumes (well under a milliliter) will give accurate results if care is taken. For manual bolus integration, the total volume of the chamber from which the sample is taken must be known, as in the techniques described above. For automatic bolus integration, the air in the chamber *is* the sample, which means that its volume does not need to be known.

Bolus integration, as the name implies, relies on integrating, or determining the area under, the O_2 or CO_2 concentration profile as the sample is swept through the system at a known flow rate (Figure 4.6). With suitable analytical techniques, the area can then be expressed in milliliters or microliters. With manual bolus integration, a subsample taken from a larger volume is usually used, in which case the volume of O_2 or CO_2 consumed or produced relative to the subsample is transformed to that figure for the chamber as a whole; thus:

$$V_x = V_m \left(V_c / V_s \right) \tag{4.23}$$

where V_x is the final $Volo_2$ or $Volco_2$, V_m is the measured $Volo_2$ or $Volco_2$ of the subsample, V_c is the volume of the chamber, and V_s is the volume of the subsample. As always, V_x, after conversion to STP if necessary, is divided by the time elapsed between sealing the chamber and taking the sample.

To integrate the sample bolus, first transform it from the measured units (typically percent O_2 and percent CO_2) into a rate of O_2 consumption and/or CO_2 production. As you will see in Chapter 9, the equations are similar to those used for flow-through respirometry. It is important to note that *these initial rates are calculated only with respect to the sample*; they are strictly tools that allow the volumes to be calculated by integration, and it is critical that they not be confused with the O_2 consumption or CO_2 production rates of the experimental organism. I refer to these whole-organism rates as Vo_2 and Vco_2, not to be confused with $Volo_2$ and $Volco_2$. To reduce the potential for confusion, I refer below to the sample "rates" as M_so_2 and M_sco_2 and to the final rates calculated for the experimental organism as Vo_2 and Vco_2.

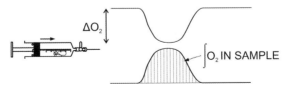

Figure 4.6 Determining O_2 volume (rather than O_2 concentration, as in Figure 4.5) by integration. A known flow rate is required. The top trace shows the original O_2 depletion signal. The bottom trace shows the same data after subtraction of F_io_2, inversion, and transformation to M_so_2 in (for example) milliliters per hour. Integrating beneath the peak against time in minutes yields milliliters of O_2 depletion in the injected sample in this example. After correction for sample versus enclosure volume (equation 4.23), this volume can be divided by enclosure duration to yield Vo_2.

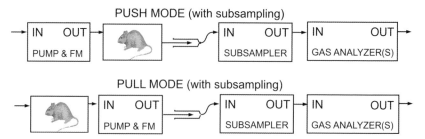

Figure 4.7 Push versus pull mode flow systems in respirometry. A combined pump and flow meter (PUMP & FM) is assumed. Subsampling of the main flow through the gas analyzers is shown; this is usually only necessary at flow rates above about 200–1000 mL min⁻¹, and is not used for manual bolus integration. Note that for clarity, scrubber columns are not shown, but they are generally necessary (see text). If subsampling is not used in the pull system, it is usually best to plumb the pump and flow meter after the analyzers.

Another important point is that *flow rates through the analyzers must be known*. In basic educational systems this can be accomplished with a rotameter, which is typically accurate to 5–10 percent. If you use a rotameter (which is a volumetric, rather than mass, flow meter), you will need to measure barometric pressure and the temperature of the rotameter and use equation 2.1 to STP-correct the measured flow rate. For research-grade systems, a mass flow meter or a mass flow controller is required. These are typically accurate to 1–2 percent or better and are already corrected to STP (see Appendix 1). Thus, they are not affected by shifts in temperature or barometric pressure.

The flow system can be configured in pull mode or push mode. Figure 4.7 illustrates the differences, with a sneak peek at flow-through respirometry; for manual bolus integration, subsampling does not take place, and the injection port plus tubing reservoirs replace the animal chamber. The equations differ slightly depending on whether a push or pull system is used, and on whether or not CO_2 is scrubbed from the air stream (and if it is scrubbed, *where* it is scrubbed), as described in Chapter 9.

I now describe how to use this technique in several different scenarios. In the first two examples, I go into some detail regarding analysis techniques, which are similar in all subsequent examples. My aim is to provide you with the most straightforward descriptions possible.

Measuring oxygen consumption only

For measuring O_2 consumption only, use the plumbing diagrammed in Figure 4.4, but with a pump and flow meter capable of delivering an accurately known flow rate. We are operating in pull mode, and absorbing water vapor and CO_2, which allows us to use the following equation, which is independent of RQ:

$$M_sO_2 = FR(F''_iO_2 - F''_eO_2)/(1 - F''_iO_2)$$ (4.24)

where M_sO_2 is the *sample* rate of O_2 consumption, calculated only for later integration to determine the volume of O_2 in the sample (as opposed to the chamber); FR is the mass flow rate (or STP-corrected volumetric flow rate); and F''_iO_2 is the fractional concentration of the water-free, CO_2-free air that provides the background or baseline O_2 concentration for the system. This is typically considered to be 0.2094 (20.94 percent, the value to which the O_2 analyzer is normally calibrated with a dry, CO_2-free ambient air stream flowing through it).

Let us go through the setup, operation, and analysis of this system step-by-step.

Create a maximally stable source of baseline air for the system. This is the air that will fill the chambers, that will be pulled through the plumbing system and the O_2 analyzer, and that will operate as the reference against which the O_2 consumption of the organism will be compared. You can control for changes in the F''_iO_2 of this air stream by running blank chambers and subtracting the readings from the chambers that contain organisms, and you should run these in any event, though if the baseline air stream is stable enough, blanks will seldom be necessary. To achieve this level of stability you should either take the air stream from outside the building, or take it from a compressed air tank or (less preferably) from a large carboy that will blunt changes in ambient F_iO_2 (Figure 4.2).

Pull the air through the analyzers, after scrubbing it of water vapor and CO_2, at a flow rate of 50–100 mL min^{-1} for a typical system.

After the system has equilibrated, span the O_2 analyzer to read 20.94 percent for dried, CO_2-free outside air. This is more accurate than any tank calibration gas you will find outside a standards institute (Tohjima et al., 2005). As Mr. T might say, I pity the poor fool who uses a 1 percent accuracy O_2 span gas while they are breathing a span gas with at least tenfold better accuracy.

Flush the chambers with the baseline air stream. You may wish to temporarily increase the flow rate for this purpose, depending on the volume of the chambers relative to the flow rate.

Seal the chambers, noting the time of sealing. It is easiest to use Luer bulkhead fittings and a Luer three-way stopcock (see Appendix 1).

Wait until a measurable O_2 depletion has been created. This can vary from a few minutes to an hour or more, depending on the organism, its size, its temperature, and the volume of the chamber. You will need to determine this by experiment, although some rough guidance can be obtained from the closed-system calculator on www.respirometry.org.

Record data from the analyzer.

Take samples from the chambers (noting the time elapsed since closure) and inject them, one at a time, into the injection port. Be sure that the tubing system can take the full volume of the injection. Wait for the analyzer to return to baseline between samples.

Save the recording.

Load the recording into (or transfer to) data analysis.

If you're recording O_2 in percent, divide the O_2 channel by 100.

Correct the baseline of the O_2 analyzer (the parts between the downward valleys corresponding to each sample injection) so that any drift is removed and the valleys trend downward from 0. This corresponds to $F''_eO_2 - F''_iO_2$.

Flip the channel by multiplying it by –1. This corresponds to $F''_iO_2 - F''_eO_2$.

Multiply the channel by the STP-corrected flow rate. If you are using a mass flow meter or controller, you are probably recording its output (transformed to flow rate) in another channel; if so, multiply by that channel.

Divide the channel by $(1 - 0.2094)$ if in your setup 0.2094 is a fair estimate of F''_iO_2. If you are using good software, all of these steps should be automatable by using a macro.

Integrate each peak (which used to be a valley prior to flipping) against time in the time units of your flow rate. For example, if your flow rate is in milliliters per minute, integrate the peak against time in minutes. The result is the volume of O_2 in the peak, in milliliters.

Thus armed with the peak volume, scale it up to the volume of the chamber (minus the animal) using equation 4.23.

Divide the scaled peak volume by the enclosure time.

Measuring both oxygen consumption and carbon dioxide production

To measure both O_2 consumption and CO_2 production, use a pull system as shown in Figure 4.8. The air stream is dried, scrubbed of CO_2, and pulled through the O_2 and the CO_2 analyzers in series. Under these circumstances, the O_2 is diluted by the CO_2 produced by the organism, so the equations are more complicated, but not intolerably so. First, calculate M_sCO_2 as follows:

$$M_sCO_2 = [FR(F'_eCO_2 - F'_iCO_2) - F'_iCO_2(M_sO_2)]/(1 - F'_iCO_2) \tag{4.25}$$

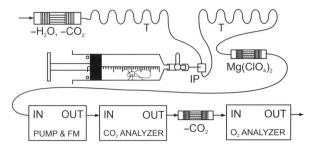

Figure 4.8 A constant volume pull mode, dynamic injection setup for both O_2 and CO_2. A known flow rate is required, supplied in this case by a combined pump and flow meter (PUMP & FM). A water vapor and CO_2 scrubber starts the flow path, into which air from the syringe containing the specimen is injected via injection port IP. For optional stability of incurrent O_2, use a carboy before the scrubber (Figure 4.2). Note that magnesium perchlorate is before the CO_2 analyzer scrubber; Drierite interacts with CO_2, slowing response times (see Chapter 19). The CO_2 scrubber is optional but simplifies the calculation of oxygen consumption rates.

where M_sco_2 is the *sample* rate of CO_2 production, calculated only for later integration to determine the volume of CO_2 in the sample (as opposed to the chamber); FR is mass flow rate (or STP-corrected volumetric flow rate); and F'_ico_2 is the fractional concentration of CO_2 in the scrubbed air that provides the background or baseline O_2 concentration for the system. M_so_2 is the sample rate of O_2 production, which in this case we do not (yet) have. How depressing. But look at F'_ico_2. In this system, it is equal to zero (or F''_ico_2) because we have scrubbed CO_2 from the air that filled the chambers and from the air that is flowing through the system. Had we not scrubbed CO_2, we would have had to know F_ico_2 and M_so_2 (although it should be pointed out that their product is still negligible unless F_ico_2 is unusually high or M_so_2 is large; you can convince yourself of this by doing a couple of quick example calculations). Because F_ico_2 is zero, though, equation 4.25 simplifies to:

$$M_sco_2 = FR(F'_eco_2 - F'_ico_2) \qquad (4.26)$$

Integrating the transformed CO_2 peak against the same time units as the flow rate yields the volume of CO_2 in the sample.

Unfortunately, analysis of the O_2 peak is not as straightforward because it must be corrected for the dilution caused by CO_2. It is far easier in practice to scrub CO_2 after the CO_2 analyzer, thus eliminating the dilution of O_2 by CO_2. In that case, you can use equation 4.23 to calculate M_so_2.

If you opt not to scrub CO_2 from the air stream, you cannot easily correct the O_2 peak for CO_2 dilution by using the fractional concentration data prior to integration, because doing so requires that the shape of the O_2 and CO_2 peaks be identical. That will seldom be the case even if the time lag between the analyzers is corrected. Instead, you can apply the CO_2 dilution correction after calculating the uncorrected O_2 volume, in two stages. First, calculate the raw O_2 rate data without dividing by the customary $(1 - F'_io_2)$:

$$M_so_2 = FR(F'_io_2 - F'_eo_2) \qquad (4.27)$$

where M_so_2 is the *sample* rate of O_2 consumption, calculated only for integration to determine the volume of O_2 in the sample (as opposed to the chamber). After this transformation, integrating under the peak (against the same time units as the flow rate) yields the raw volume of O_2 depleted from the sample. It is now straightforward to apply the CO_2 dilution correction and complete the calculation of the O_2 depletion volume:

$$Volo_2 = [VoluO_2 - F'_io_2(Volco_2)]/(1 - F'_io_2) \qquad (4.28)$$

where $Volo_2$ is the corrected volume of O_2 depletion, $VoluO_2$ is the uncorrected volume of O_2 depletion, and $Volco_2$ is the sample volume of CO_2. As you can see, it is far easier to simply scrub the CO_2 prior to O_2 analysis and use equation 4.23. The sample volumes are then scaled and divided by enclosure duration to yield Vo_2 and Vco_2.

Let us go through the setup, operation, and analysis of this system step by step.

Create a maximally stable source of baseline air for the system, as described for the O_2-only system. Be sure to scrub CO_2 from this air in addition to water vapor.

Pull the air through the analyzers at a flow rate of 50–100 mL min^{-1} for a typical system (Figure 4.8). Be sure to include a water vapor scrubber so that any water vapor in the injected sample is removed.

After the system has equilibrated, span the O_2 analyzer to read 20.94 percent for dried, CO_2-free outside air (Tohjima et al., 2005). Zero the CO_2 analyzer (remember that the baseline air stream flowing through the system has been scrubbed of CO_2).

Flush the chambers with the baseline air stream. You may want to temporarily increase the flow rate for this purpose, depending on the volume of the chambers relative to the flow rate.

Seal the chambers, noting the time of sealing. It is easiest to use Luer fittings and a Luer three-way stopcock (see Appendix 1).

Wait until a measurable O_2 depletion and CO_2 enrichment have been created. This can vary from a few minutes to an hour or more, depending on the organism, its size, its temperature, and the volume of the chamber. You will need to determine this by experiment, although some rough guidance can be obtained from the closed-system calculator on www.respirometry.org.

Record data from the analyzers.

Take samples from the chambers (noting the time elapsed since closure), and inject them, one at a time, into the injection port. Be sure that the tubing system can take the full volume of the injection. Wait for the analyzers to return to baseline between samples (F_iO_2 for the O_2 analyzer, zero for the CO_2 analyzer).

Save the recording.

Load the recording into (or transfer to) data analysis.

Divide the CO_2 channel by 100 if you recorded CO_2 in percent or by 10^6 if you recorded CO_2 in parts per million.

Correct the baseline of the CO_2 analyzer (the parts between the peaks corresponding to each sample injection) so that any drift is removed and the peaks trend upward from 0. This corresponds to $F_eCO_2 - F_iCO_2$, where F_iCO_2 is zero, so really, you're just removing any residual analyzer drift.

Multiply the CO_2 channel by STP-corrected flow rate. If you are using a mass flow meter or controller, you are probably recording its output (transformed to flow rate) in another channel; if so, multiply by that channel. You now have M_sCO_2.

Integrate under the M_sCO_2 peak to yield the sample volume of CO_2, VolCO$_2$.

Select the O_2 channel. If you're recording O_2 in percent, divide the channel by 100.

Correct the baseline of the O_2 analyzer (the parts between the downward valleys corresponding to each sample injection) so that any drift is removed and the valleys trend downward from 0. This corresponds to $F_eO_2 - F_iO_2$.

Flip the O_2 channel by multiplying it by –1. This corresponds to $F_iO_2 - F_eO_2$.

Multiply the O_2 channel by STP-corrected flow rate. If you are using a mass flow meter or controller, you are probably recording its output (transformed to flow rate) in another channel; if so, multiply by that channel.

If you are scrubbing CO_2 prior to O_2 analysis (recommended), divide the result by $(1-0.2094)$, if in your setup 0.2094 is a fair estimate of F''_iO_2, to yield M_sO_2.

Integrate under the M_sO_2 peak to yield the sample volume of O_2 depletion, $VolO_2$ (if you are scrubbing CO_2 prior to O_2 analysis), or the uncorrected sample volume of O_2 depletion, $VoluO_2$ (if you are not.)

If you are not scrubbing CO_2 prior to O_2 analysis, subtract $Volco_2(0.2094)$ from $VoluO_2$, if in your setup 0.2094 is a fair estimate of F''_iO_2. Then divide the result by $(1-0.2094)$ to yield $VolO_2$.

Thus armed with the peak O_2 and CO_2 volumes, scale them up to the volume of the chamber (minus the animal) using equation 4.23 and divide by enclosure time to get rates. If you are using good software, all of these steps should be automatable by using a macro.

It is worth emphasizing the very different ways in which O_2 and CO_2 are measured in these systems. CO_2 is measured against a zero baseline if, as recommended, CO_2 is scrubbed from the baseline air. Thus, any CO_2 injected into the system will yield an upward-going signal. O_2, in contrast, is measured *relative to baseline air*; in other words, only samples that differ in O_2 concentration from the baseline will be detected. Subtracting the baseline is then equivalent to subtracting F_iO_2 from F_eO_2. Because the equations expect $(F_iO_2 - F_eO_2)$, you must multiply the result by -1, effectively reordering the subtraction.

Measuring only carbon dioxide production

To measure only CO_2 production, use a pull system as shown in Figure 4.8, but without the O_2 analyzer. The air stream is dried and pulled through the CO_2 analyzer. Under these circumstances, the CO_2 is slightly enriched because O_2 is consumed by the organism. Because you are not measuring O_2, you don't know the magnitude of this (admittedly rather minor) effect. So, you need to guess at its magnitude by assigning an RQ (Vco_2/Vo_2). Then, calculate $Msco_2$ as follows:

$$M_sco_2 = FR(F'_eco_2 - F'_ico_2)/\{1 - F'_ico_2[1 - (1/RQ)]\} \tag{4.29}$$

where M_sco_2 is the *sample* rate of CO_2 production, calculated only for later integration to determine the volume of CO_2 in the sample (as opposed to the chamber). Because F_ico_2 is equal to zero because we have scrubbed CO_2 from the air that filled the chambers and from the air that is flowing through the system, equation 4.29 simplifies to:

$$M_sco_2 = FR(F'_eco_2 - F'_ico_2) \tag{4.30}$$

Let us go through the setup, operation, and analysis of this system step by step.

A particularly stable source of baseline air is unnecessary if, as recommended, you scrub CO_2 from the baseline air in addition to water vapor.

Pull the air through the analyzer at a flow rate of 50–100 mLmin^{-1} for a typical system (Figure 4.8). Be sure to include a magnesium perchlorate water vapor scrubber so that any water vapor in the injected sample is removed. Unlike Drierite or silica gel, magnesium perchlorate does not interact with CO_2.

After the system has equilibrated, zero the CO_2 analyzer (remember that the baseline air stream flowing through the system has, optimally, been scrubbed of CO_2).

Flush the chambers with the baseline air stream. You may want to temporarily increase the flow rate for this purpose, depending on the volume of the chambers relative to the flow rate.

Seal the chambers, noting the time of sealing. It is easiest to use Luer taper fittings and a Luer three-way stopcock (see Appendix 1).

Wait until a measurable CO_2 enrichment has been created.

Take samples from the chambers (noting the time elapsed since closure), and inject them, one at a time, into the injection port. Be sure that the tubing system can take the full volume of the injection. Wait for the analyzer to return to zero between samples.

Save the recording.

Load the recording into (or transfer to) data analysis.

Divide the CO_2 channel by 100 if you were recording CO_2 in percent or by 10^6 if you were recording CO_2 in parts per million.

Correct the baseline of the CO_2 analyzer (the parts between the peaks corresponding to each sample injection) so that any drift is removed and the peaks trend upward from 0. This corresponds to $F_e\text{co}_2 - F_i\text{co}_2$, where $F_i\text{co}_2$ is zero, so really, you're just removing any residual analyzer drift.

Multiply the CO_2 channel by STP-corrected flow rate. If you are using a mass flow meter or controller, you are probably recording its output (transformed to flow rate) in another channel; if so, multiply by that channel. You now have $M_s\text{co}_2$.

Integrate each CO_2 peak against time in the time units of your flow rate. For example, if your flow rate is in milliliters per minute, integrate the peak against time in minutes. The result is the volume of CO_2 in the peak in milliliters.

Scale the CO_2 volumes up to the volume of the chamber (minus the animal) using equation 4.23 and divide by enclosure time to get the CO_2 emission rate.

Analyzing small gas samples

In some cases, you might want to analyze a small sample of gas and determine its volume of O_2 and/or CO_2 directly, often for reasons that have nothing to do with respirometry. You could measure O_2 relative to background O_2 concentration, but it is more direct in such a case to inject the sample into an O_2-free gas stream. In this case you are measuring not only CO_2 but O_2 in absolute terms. A minor modification to the system described above will accomplish this. Simply flow nitrogen through the system to establish a zero-O_2 and zero-CO_2 baseline and integrate under the resulting peaks as described above after correcting for analyzer drift (without, of course, flipping the O_2 trace), bearing in mind that $F_i\text{o}_2$ is zero. This technique works best with O_2 analyzers that are designed to operate optimally near zero O_2; zirconia cell and paramagnetic O_2 analyzers tend to work better against a zero-O_2 baseline than fuel cell analyzers. Tiny sample volumes (from a couple of milliliters to a few nanoliters if you are using good analyzers) are sufficient. One example of this technique is described by Goldstein et al. (1991). The source of nitrogen can be a manifold (an empty syringe barrel works well; see Figure 16.1) into which bottled N_2 flows at a rate at least double the rate from which the system extracts it.

Automatic bolus integration

For automatic bolus integration, take the following steps (we are considering, for now, only a single chamber).

Flush the chamber with baseline air—optimally air scrubbed of CO_2. It is critical that the baseline air O_2 concentration is stable if you are measuring O_2.

Seal the chamber for a period of minutes to hours, depending on its volume and the metabolic activity of your specimen.

During the time that the chamber is sealed, the baseline air stream flows through the gas analyzers, establishing the system's zero (CO_2) or F_iO_2 (O_2).

Flush the chamber through a water vapor scrubber and then through O_2 and/or CO_2 analyzers, through which baseline air normally flows.

Calculate the volumes of O_2 consumed and/or CO_2 produced by integration under the O_2 depletion and CO_2 enrichment boli, as described in the section "Analyzing small gas samples," then divide by chamber enclosure duration to obtain rates.

This technique has an extremely advantageous feature: It is not necessary to know the chamber volume. This is because all of the chamber's air is replaced, and its volumes of O_2 depletion and CO_2 enrichment are directly measured, so the original volume of the chamber does not matter. This is only true within certain limits, however; care should be taken to provide a large enough chamber so that the O_2 concentration does not fall, and the CO_2 concentration does not rise, to levels within the chamber that may affect the organism being measured. Keeping changes in either gas to < 1 percent over the course of containment will usually not stress the specimen.

It is certainly possible to use this technique manually by moving tubing connectors or operating valves by hand rather than electrically controlling the switching of chambers, but it is also straightforward to automate it using gas solenoids or gas stream multiplexers (see Appendix 1) if you are using a good computerized data acquisition and analysis system with instrument control capability. So equipped, this technique becomes a formidable metabolic data acquisition engine for small to medium specimens (depending on container volume) of almost any kind.

Figure 4.9 shows this principle. Where multiple chambers are used, the steps are as follows.

An air stream of stable composition and a known mass flow rate, typically scrubbed of water vapor and CO_2, is pulled through one or more gas stream multiplexers that allow it to pass through one (and only one) of N chambers. After passing through the selected chamber, the air stream is pulled through a water vapor scrubber and then through one or more gas analyzers, typically O_2 and CO_2 (sometimes CH_4 as well).

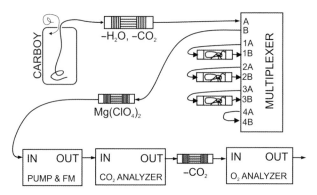

Figure 4.9 A constant volume pull mode, automatable constant volume respirometry system incorporating both O_2 and CO_2 analysis. Note that chamber volumes do not need to be known. A three-animal system is shown; systems having up to 60+ channels are easily configured. The multiplexer (four-channel unit shown) is controlled by a computer running appropriate software, which sequentially flushes the chambers and also handles data analysis. The CO_2 scrubber before the O_2 analyzer is optional. One channel (number 4 in this case) is reserved for use as a baseline channel for measuring F_iO_2 and F_iCO_2. A carboy (see also Figure 4.2) stabilizes incurrent F_iO_2.

At the beginning, seal all chambers, except for one that is empty and acts as the system zero, through which the baseline air stream flows when it is not directed elsewhere. In the most common configuration, this means that a pair of solenoid air valves to either side of the baseline chamber is open, while similar pairs to either side of each other chamber in the system are closed. I arbitrarily call the baseline chamber 0.

A data acquisition system records data from the gas analyzers and automatically directs the opening and closing sequences described below.

The next chamber (chamber 1) is opened and flushed for a set time, typically a minute or two, depending on chamber volume and flow rate. The system selects the time to allow the O_2 and CO_2 concentrations in the air leaving the chamber to return to baseline levels; optionally, you should monitor the gas analyzers to ensure this is the case.

The system closes that chamber and records the time. The next chamber is opened and likewise flushed, then it is sealed and the time is noted.

This procedure is repeated until some maximum chamber number is reached, or until there are no more chambers. The system can optionally return to the baseline chamber between flushing a given number of specimen chambers for optimal accuracy.

Chamber 0 (the baseline) is again opened, and the system monitors baseline air. All specimen chambers are now sealed, and the creatures within them are consuming O_2 and producing CO_2. This step can last a variable length of time, typically set by an operator depending on the expected gas exchange rates of the samples. The goal is usually a compromise between minimizing the enclosure duration and obtaining a usable O_2 depletion and CO_2 emission signal. In setups with many chambers, this step can often be eliminated because the cycle time for reading each chamber is long enough for usable changes in O_2 and CO_2 concentrations to develop without requiring an additional pause.

The system then sequentially opens and flushes each chamber, with (optionally) a return to baseline levels between chambers. As each bolus is swept through the gas analysis system, it is recorded. The time at the end of the chamber flush is likewise recorded.

These procedures are repeated as often as required to sample from all of the chambers, or to resample the chambers periodically to create a time-course.

You can analyze the recorded data in real time or (preferably, in most cases) after the run is complete. During analysis, the O_2 and CO_2 data are baseline corrected, converted to rates, integrated, and the resulting volumes of O_2 and CO_2 are calculated. If CH_4 is monitored, it can be treated analogously to CO_2. The volumes are then divided by the interval between successive sealings of each chamber to yield rates.

Automated bolus integration systems are available as turn-key products from a few vendors (see www.respirometry.org), but can be created from scratch by anyone with above-average technical and computer expertise who is willing to spend several weeks on the project and make painful sacrifices in his or her personal life.

USING MOIST AIR TO FILL CHAMBERS

Some samples are prone to desiccation if they are exposed to dry air. Examples include soil samples and many invertebrates. The systems described work best with dry air, but there are ways to raise the relative humidity of the air stream that provides the chambers with air (for example, by bubbling it through water) so that stress on the experimental samples is minimized. This is quite easy to achieve if only O_2 analysis is contemplated. With CO_2 analysis that requires scrubbing of CO_2 from the incurrent air, the problems are that CO_2 absorbents interact with water vapor, and that liquid water is a mighty capacitor as far as CO_2 is concerned. In general, the best strategy is first to scrub water vapor and CO_2 from the air, and then add water vapor back into the air stream. This subject is explored in depth below.

Whether O_2 or CO_2 analysis is planned, there will be another problem to deal with. As we saw in the detailed treatment of the static injection technique, O_2 in particular is subject to huge errors caused by water vapor dilution. The added water vapor will have to be either scrubbed or measured before analysis and compensated for. This latter approach is easy in theory, but tricky in practice where gas signals vary rapidly, as occurs here when an O_2-depleted gas bolus passes through the system. For more details on dilution compensation, see the overview chapter on flow-through respirometry (Chapter 8 and especially equation 8.7). In this case, it's definitely easier to scrub water vapor than to compensate for its presence.

Here are some suggestions for adding water vapor to air during respirometry and removing its influence before or during analysis. I deal in turn with obtaining a supply of CO_2-free air; adding water vapor; and removing water vapor before gas analysis (see "Option A" for measuring it and compensating for its presence).

If you are analyzing CO_2, you will generally need to scrub it from the air before adding water vapor. Even if you are working only with O_2, scrubbing CO_2 from incurrent air simplifies the equations you will use, as described above. A large Drierite/Ascarite/Drierite column (see Chapter 19) will work well for low flow rates (50–200 mLmin^{-1}), but it will become quite rapidly depleted at higher flow rates or in moist or CO_2-rich surroundings. You may want to prescrub the air entering the column with cheaper chemicals, such as silica gel (which can be regenerated gracefully) for water vapor and granular soda lime for CO_2.

A better long-term solution is to use an automatic regenerating scrubber that removes both water vapor and CO_2 (see Chapter 19). Any residual traces of water vapor and CO_2 can be removed by a scrubber column, which will last almost indefinitely. Regenerating scrubbers are, unfortunately, quite expensive and require a source of high-pressure compressed air. In practice, the output of the regenerating scrubber is best sent directly into a large carboy or tank, which will damp the inevitable fluctuations in the O_2 concentration of the compressed air supply (Figure 4.2). This step is critical for precise O_2 analysis. Air can then be pulled from the tank, residual water vapor and CO_2 scrubbed, and the air stream rehumidified.

An alternative approach is to use an aqueous CO_2 scrubber with a known water vapor pressure. A KOH solution of a known concentration will, in addition to scrubbing CO_2, help define the relative humidity of the air (see equation 2.6). If you use this approach, be careful to trap caustic aerosols or particulates downstream from the bubbler. Avoid this approach if you aren't trained and completely comfortable working with large containers of highly caustic chemicals. Also, be aware that the KOH solution will pick up water vapor from incoming air that is moister than the solution's equilibrium humidity, and it will lose water if the converse is true. You will need to monitor the solution and amend or replace it when necessary.

To add water vapor back into air, the easiest and most obvious approach is to use a water bubbler (also called a gas washer). You will need to run the bubbler for some time (up to several hours, depending on its size) before it completely equilibrates with incoming gas concentrations; it is usually best to leave it running for this reason. However, if you do not scrub the water prior to analysis, at all costs avoid condensation in the tubing, respirometer chamber, and especially the gas analyzers. In fact, zirconia-based O_2 analyzers can be, and often are, ruined by a steam-driven sensor explosion if they take up liquid water (Chapter 16). The explosion is not dramatic, but its effect on the lab budget can be. In addition, liquid water in chambers and tubing acts as a gas sink, especially for CO_2, and it should be avoided for that reason, too. Using a gas bubbler at room temperature is therefore a risky activity. There are two ways around this problem of controlling the water vapor content of the air entering the respirometry system.

In the first technique, chill the bubbler bottle to some defined temperature a few degrees below room temperature. This can be accomplished with a cooling water bath, a chill plate, or a small temperature-controlled cabinet. This drops the water vapor pressure of the output air to a level corresponding to the dewpoint of the bubbler bottle, so it will not condense at room temperature. This also sets the relative humidity of the air, which is simply the water vapor pressure of the air divided by the saturated water vapor pressure at room temperature, expressed as a percentage. For example, at a room temperature of 25°C, the saturated water vapor pressure is 3.2 kPa (a water vapor pressure calculator can be found at www.respirometry.org). If you have chilled the bubbler down to 15°C, the water vapor pressure of the resulting air stream will be 1.7 kPa, so the relative humidity of the air stream is 100(1.7/3.2), or 53 percent. An accurate thermocouple probe in the water can be used to monitor the bubbler's temperature.

An alternative, innovative variant of this technique passes the incurrent air stream through tubing made of Nafion, a polymer permeable to water vapor but not to O_2 or CO_2, which is immersed in water at a known temperature (see Chapter 19 for more on Nafion and its uses). The Nafion tubing eliminates problems with splashing and with interaction between the gas stream and CO_2 dissolved in the water. It is also remarkably effective; an ~ 20-cm length of submerged Nafion will hydrate air efficiently to the dewpoint of the water at modest flow rates of ~ 100 mL min^{-1}. An online calculation for converting the dewpoint to water vapor pressure can be found at www.respirometry.org; see also equation 16.7.

In the second technique, a condenser at a controlled temperature controls the dewpoint, and thus the water vapor pressure, of the air after the bubbler by condensing excess water (Figure 4.10). The condenser can be a glass type with a water-jacket fed by a cooling water bath, or a bench-top device with an active heat pump such as a Peltier effect device. In either case the condensate is periodically removed by draining or pumping;

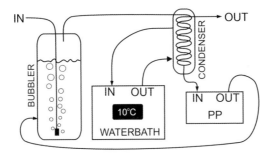

Figure 4.10 Controlling the dewpoint, and thus the water vapor pressure, of an air stream by condensing excess water vapor. At a constant temperature of the air stream at the target location, this approach also controls the relative humidity of the air stream. A bubbler raises the water vapor pressure of the input air stream to near ambient levels (e.g., 20°C). A water bath maintains a water-jacketed condenser at 10°C in this example (commercial dewpoint controllers often use Peltier effect heat pumps and metal condenser plates). Condensate is pumped out of the condenser at a slow rate by peristaltic pump (PP), which optionally recycles the condensed water back into the bubbler. Air exits the system at a dewpoint of 10°C (1.22 kPa water vapor pressure). At 20°C the saturated water vapor pressure is 2.33 kPa, so the resulting RH is 100(1.22/2.33) = 52.4 percent. An online calculator for converting the dewpoint to water vapor pressure can be found at www.respirometry.org.

a spare peristaltic pump works well. As with the previous technique, the relative humidity of the air stream can be accurately defined. Some bench-top condensers will even allow the operator to set a relative humidity and will actively regulate that relative humidity by adjusting the condenser's dewpoint while monitoring room or chamber temperature. This also allows relative humidity to be changed flexibly during experiments, if needed.

Turning now to the task of removing water vapor from the humidified air stream before analysis, several distinct techniques can be used (see Chapter 19. Scrubber chemicals can certainly be used, but they require frequent regenerating. In the case of CO_2 analysis, it must also be borne in mind that Drierite slows down response speed unacceptably because it adsorbs CO_2. Magnesium perchlorate must be used instead. It is usually best, and also more economical in the long run, to remove most of the water vapor with a nonexpendable dryer and then to use a chemical scrubber to remove the last traces of water vapor. Two common types of nonexpendable gas sample dryers are chilling condensers, which chill the sample to about 1–5 °C and condense excess (but not all!) water vapor, and dryers that use Nafion tubing and a reflux system to accomplish the task (see Chapter 19).

5

Aquatic Oxygen Analysis

Four different techniques can be used to determine O_2 consumption rates of aquatic organisms. One is coulometry (Chapter 3), which is probably the best technique (Hoegh-Guldberg and Manahan, 1995). Unfortunately, off-the-shelf coulometric equipment does not exist at present. Another technique, coming into increasing use, is headspace analysis; the aquatic organism is kept in a layer of water in a sealed chamber with a headspace containing air. The headspace air is periodically swept out, its composition analyzed, and the air is then replaced with fresh air of known composition. This is basically automatic bolus integration, described in Chapter 4. It is also possible to implement a flow-through variant of this technique if sufficiently sensitive analyzers are used. Alternatively, the water may be circulated through a membrane array with a large surface area from which gas signals can be extracted. I will deal with headspace analysis below. The third, still-common technique uses a polarographic O_2 electrode, commonly known as a Clark electrode, which was invented by Leland Clark (Clark et al., 1953). A fourth technique, fiber-optic-based O_2 analysis that relies on fluorescence quenching, is now firmly in ascendance. The latest fluorescence-based designs, the best of which rely on the decay kinetics of fluorescence quenching rather than its absolute magnitude, are ready for prime time. You can use them like Clark electrodes, but because they don't consume O_2, they are easier and more forgiving to work with. The numerous warnings about boundary layer effects and annoying chemical quirks that apply to Clark electrodes (see Appendix 2) do not apply to them. The analytical calculations and other details that apply to both Clark and fluorescence-based probes are described below. To minimize clutter in this chapter, I discuss Clark electrodes in detail in Appendix 2. Bear in mind, however, that Clark electrodes may still be required if very high resolution or operation at very low O_2 tensions is needed in your application.

CALIBRATING O_2 ELECTRODES

O_2 electrodes and their accompanying analyzers require frequent calibration. A two-point calibration is typical, with one point being at nominal zero po_2 and the other at saturated po_2 under the conditions of the experiment. The zero po_2 solution is usually 10–50 mmol sodium dithionite solution, but N_2-bubbled water can also be used. The O_2-saturated solution is

Measuring Metabolic Rates: A Manual for Scientists. Second Edition. John R. B. Lighton, Oxford University Press (2019). © John R. B. Lighton (2019). DOI: 10.1093/oso/9780198830399.001.0001

usually prepared by bubbling air through it. This approach must be used with caution. Because water evaporates during bubbling, the latent heat of vaporization can cause the temperature of the water to drop. When the bubbling stops and the water temperature rises, O_2 super-saturation can result. In addition, the bubbling exposes the water to air at slightly above atmospheric pressure (or no bubbles would be produced), with the same result. As an alternative, the O_2 content of air saturated with water vapor can be used. Because the response of most O_2 electrodes is linear versus po_2, intermediate points are usually not required.

This brings us to what "100 percent O_2 saturation" means in terms of the O_2 content of the solution. This is where things can get complicated. Henry's law states that the partial pressure of a gas in solution is proportional to its partial pressure in a gas mixture in contact with the solution. Thus, the amount of O_2 dissolved in water drops as barometric pressure decreases, so you need to know the local barometric pressure for the calibration to be meaningful. This is not a trivial adjustment; Denver, Colorado and Furnace Creek in Death Valley, California differ in this respect by about 20 percent. In addition, the amount of O_2 held in solution at saturation is a strong function of temperature. An empirical equation can be used to predict the saturated O_2 content of pure water at a barometric pressure of 101.3 kPa:

$$\text{mg } O_2 L^{-1} = 0.0035T^2 - 0.336T + 14.4 \tag{5.1}$$

where T is the water temperature in degrees Celsius. When saturated O_2 content at standard pressure is calculated using equation 5.1, you can calculate the actual saturated O_2 content at the local barometric pressure:

$$\text{mg } O_2 L^{-1} = \text{BP}(\text{mg } O_2 L^{-1}) / (101.325) \tag{5.2}$$

where BP is local barometric pressure in kilopascals. This is the figure to which you should set the O_2 analyzer with the electrode held in stirred, O_2-saturated pure water or held in water-vapor-saturated air. Note the "pure" water; if you are using seawater, and indeed if the water contains any significant dissolved species apart from O_2, the less O_2 can dissolve in it, and the calibration will have to make allowance for this. The degree to which this holds is also a function of temperature. For example, at standard pressure and 20°C, fresh water holds 9.1 mg O_2 L^{-1} at saturation, while in seawater this drops to about 7.5 mg O_2 L^{-1}. All of this makes calibration of O_2 electrodes in water-vapor-saturated air quite attractive.

Tables of saturated O_2 content versus solute concentrations and temperatures can be found at www.respirometry.org. The concentration of O_2 in the water can also be expressed in other units, notably parts per million, which are exactly equivalent to milligrams per liter. Milligrams of O_2 can easily be converted to volume (1 mg O_2 = 0.70 mL O_2 at STP) or molar (1 mg O_2 = 31.25 mmol) units.

The two most common forms of aquatic respirometry are constant volume and flow-through respirometry. These are closely analogous to their aerial cousins. I consider them separately, then turn to headspace analysis.

CONSTANT VOLUME AQUATIC RESPIROMETRY

Constant volume aquatic respirometry is straightforward. Trap your specimen in a leak-proof container with no bubbles, into which your O_2 electrode is inserted via an airtight seal, and which contains a small stirrer bar if your specimen isn't consistently active (Figure 5.1). The electrode reads the O_2 concentration within this container.

Knowing the volume of the container minus the animal and stirrer bar (see Chapter 4 on constant volume and constant pressure respirometry for methods of estimating chamber volumes), you can, while acquiring data, multiply the measured concentration in milligrams O_2 per liter by the volume of the container in liters, thus obtaining the total milligrams O_2 in the container. This will decline over time as the specimen consumes O_2 (Figure 5.2).

From the rate of the decline in time units of your choice, determined by regression analysis in your data acquisition and analysis package, you obtain the rate of O_2 consumption—for example, in micromoles per minute or milligrams per hour. Thus,

$$Vo_2 = \Delta O_2(V) \tag{5.3}$$

where ΔO_2 is the slope of the regression relating O_2 concentration to time (for example, in milligrams per liter per minute), V is the volume of the container in the same volume units (e.g., liters), and Vo_2 is the rate of O_2 consumption in the same time units as the slope. Using Figure 5.2 as an example, the slope is -0.01396 mg O_2 L^{-1} min^{-1}. If the container held 2.0 L of water, the organism's rate of O_2 consumption would be 2.792 mg O_2 min^{-1}.

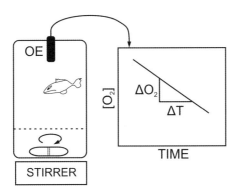

Figure 5.1 A typical closed aquatic respirometry system. The oxygen electrode (OE) is sealed inside a closed vessel containing water and the experimental animal. The vessel should be kept at a constant temperature and no air bubbles should be present. If you are using a Clark electrode, a magnetic stirrer circulates the water to break up boundary layers at the electrode's membrane and to ensure adequate mixing of the water. A grid prevents the animal from contacting the stirrer. The rate of decline of the measured O_2 concentration in the water, in combination with the volume of the vessel, yields the rate of O_2 consumption of the animal.

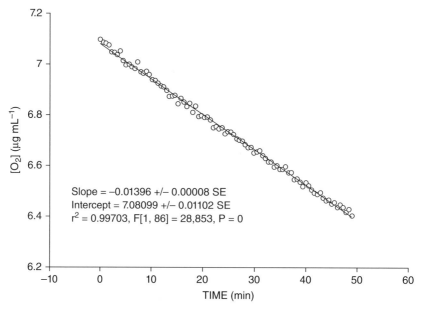

Figure 5.2 Typical graph of declining O_2 concentration within a closed aquatic respirometry system (Hoegh-Guldberg, unpublished data from the Lizard Island field site, Great Barrier Reef, Australia).

A wide variety of containers can be used. These range in sophistication from a peanut-butter jar or a preserve bottle with an O_2 electrode inserted through the lid via a grommet, to beautiful water-jacketed masterpieces worthy of a Murano glassblower. The volume can range from a liter or more to less than 1 mL for mitochondrial work. If air bubbles aren't present to act as O_2 capacitors, and if stirring takes place (Clark electrodes only!) and water temperature is constant, all will work well.

FLOW-THROUGH AQUATIC RESPIROMETRY

In flow-through aquatic respirometry, ideally we use two electrodes, one for the incurrent and one for the excurrent water stream (Figure 5.3). Both electrodes are zeroed and spanned as described. The chamber containing the specimen is fed from a holding tank of stable, O_2-saturated water pumped through the respirometer chamber at a known flow rate. Often, the water leaving the chamber is recycled into a holding tank to be reoxygenated.

The two electrodes, one before and one after the chamber, connected to their respective analyzers (or channels of a multichannel analyzer), are recorded and their outputs subtracted and multiplied by flow rate:

$$Vo_2 = (C_i O_2 - C_e O_2)\text{FR} \qquad (5.4)$$

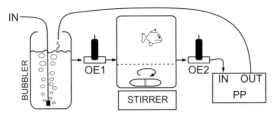

Figure 5.3 A simple flow-through aquatic respirometry system. A pump (in this example a peristaltic pump, PP) pulls O_2-saturated water past an O_2 electrode (OE1) that measures incurrent O_2 concentration, past an experimental animal (the stirrer is optional and can be omitted if the flow rate is high enough or you are using fluorescence-based O_2 probes), and past a second oxygen electrode (OE2) that measures O_2 concentration. The O_2-depleted water is optionally recycled to the bubbler that maintains O_2 saturation.

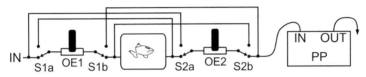

Figure 5.4 A flow-through aquatic respirometry system that implements crossover switching between incurrent and excurrent water streams. Plumbing lines that cross over without a connection circle are not connected. All four solenoid valves are energized at once. When the solenoid valves are not energized, incurrent water flows through S1a, then the first O_2 electrode (OE1), then S1b, and into the chamber. Flow leaves the chamber via S2a, the second O_2 electrode (OE2), S2b, and enters the peristaltic pump (PP). When the solenoid valves are energized, incurrent water flows through S2a, then OE2, then S2b, and into the chamber. Flow leaves the chamber via S1a, OE1, then S1b, and enters the PP. This allows one O_2 electrode to be calibrated against O_2-saturated water (bubbler not shown), while the other electrode (which was recently freshly calibrated) measures excurrent O_2 concentrations. Interelectrode drift is practically eliminated and flow direction is maintained in all sections. This method is used in Erich Gnaiger's Ouroboros aquatic O_2 analyzers and for ultra-precise atmospheric O_2 analysis (Stephens et al., 2007).

where Vo_2 is the rate of O_2 consumption in milligrams per minute, C_io_2 and C_eo_2 are the incurrent and excurrent O_2 concentrations in milligrams per liter, respectively, and FR is the flow rate of water through the system in liters per minute (obviously, you can use other units if they are compatible). Flow rate, if it is stable, can be measured by weighing the amount of water added to a container (or by looking at the graduations on a graduated cylinder) over a known interval. If the water stream does not provide enough mixing for stable O_2 readings, you will need to house a magnetic flea stirrer below each O_2 electrode or angle the electrodes so that the flow directly impinges on them. You could conceivably manage with only a downstream electrode, but consider the discussion below.

With a little ingenuity, this system can be substantially improved, as my friend Erich Gnaiger of the University of Innsbruck realized some years ago. Simply arrange for the upstream and downstream O_2 electrodes to be swapped on a regular basis (Figure 5.4).

What was the excurrent electrode is now the incurrent electrode, and vice versa. This means that any electrode calibration drift can be eliminated in software because each electrode can be periodically recalibrated with O_2-saturated water.

HEADSPACE ANALYSIS

Headspace analysis in its simplest form is simply a variant of stop-flow analysis (Chapter 4) or flow-through analysis (Chapter 8) conducted by pushing or pulling air from above the water layer in a sealed container. See those chapters and Pfeiffer et al. (2011) for more details.

A more novel variant of headspace analysis (Harter et al., 2017) utilizes a hollow fiber membrane (HFM) with a large surface area, made of gas-permeable hollow fibers of polymethylpentene (see Appendix 1) that permit relatively free diffusion of O_2 and CO_2. Harter and colleagues arranged 34 such fibers in parallel within a hollow tube through which they perfused water driven by a peristaltic pump from a closed system containing a dragonfly nymph. They then pushed dry, CO_2-free air through the parallel fibers and into a sensitive CO_2 analyzer. The result was a remarkably sensitive and simple metabolic measurement system that can, in principle, be adapted to work with a wide variety of aquatic organisms.

TRACEABILITY

Whatever modality of aquatic O_2 analysis you choose, for research purposes it is essential (not just optional) to record your data in raw units, such as O_2 concentration and flow rate, and to calculate final results (e.g., Vo_2) from the raw units after the raw units have been saved. This allows low-level problems such as electrode drift to be diagnosed and corrected and provides you with an audit trail for your data. Trust the data from any system, aquatic or otherwise, that only gives you the final results as much as you trust the edibility of a handful of half-eaten French fries lying on the sidewalk. You never know where they (the data or the fries) have been.

6

Direct Calorimetry

All metabolism equates, in the final analysis, to heat production. A candle produces heat at a rate of about 75 J s^{-1}, or 75 W (see Hamins et al., 2005, for an in-depth look at candle flames, another field pioneered by Michael Faraday). A typical human being produces much the same amount of heat, though of course over a much larger area, which is why humans, though warm and wonderful, make inferior tea candles. Calorimetry, which derives from the Latin *calor* (heat) and the Greek *metrion* (measure), can measure the heat produced by a candle (or a human, or an ant). There are three major species of calorimetry: bomb calorimetry, indirect calorimetry, and direct calorimetry.

BOMB CALORIMETRY

Bomb calorimetry sounds rather alarming, and can be so in practice too. In essence, a known mass of sample is fully combined with O$_2$, and the resulting heat production is measured. Care must be taken to ensure that enough O$_2$ is present to oxidize the sample completely, so combustion of the sample takes place in pure O$_2$ at a high pressure, typically about 3 mPa. The combustion chamber is usually made of thick stainless steel divided into two hemispheres that screw or bolt together. The structural integrity of this contraption is important. "Bomb," though nominally referring to the explosive combustion of the sample, can also refer to the effects of inadequate pressure containment. Sloppy workers do not last long in this field. Ignition of the sample typically takes place via a spark or heated wire filament. The heat produced by combustion is quantified via the resulting rise of temperature, measured either from the calorimeter walls or in a stirred water bath within which the calorimeter is immersed.

To convert the temperature change to the energy output of the combusted sample, the bomb calorimeter must first be calibrated with samples of known mass and known heat of combustion. Traditionally the calibration is carried out with benzoic acid, with a heat of combustion of 26.43 kJ g^{-1}. Bomb calorimeters differ widely in the algorithms by which the temperature change is converted into the energy content of the sample, and the manufacturer's recommendations should be consulted and used as a starting point, at least. Returning to our candle, we find, using benzoic acid to calibrate our bomb

Measuring Metabolic Rates: A Manual for Scientists. Second Edition. John R. B. Lighton, Oxford University Press (2019).
© John R. B. Lighton (2019). DOI: 10.1093/oso/9780198830399.001.0001

calorimeter, that a certain type of paraffin wax has a heat of combustion of 43.1 kJ g^{-1}. Placing a candle made from this wax on a balance shows us that it is burning paraffin wax at a rate of 102.6 mg min^{-1}, from which we determine that it is producing heat at a rate of 43,100 $\times$ 0.1026/60, or 70 J s^{-1}. As described in Chapter 14, this technique, typically carried out not with a paraffin wax candle but an absolute ethanol lamp or a propane burner, is a "gold standard" for calibrating respirometry systems.

In the same fashion, we can determine the metabolic rate of an animal not by burning it, but by quantifying the total energy content of the food entering its body minus the energy content of the urine and feces leaving it. For example, a lab rat might eat 22.4 g day^{-1} of rat chow having a heat of combustion (measured using a bomb calorimeter) of 12.3 kJ g^{-1}. Thus, its energy input is 22.4 $\times$ 12.3 or 275.5 kJ day^{-1}. Bomb calorimetry shows that its dried feces and urine account for 48.8 percent of this figure, so the rat's metabolic rate is 275.5 (1 – 0.488), or 141.1 kJ day^{-1} or 1.63 W. Alert readers will notice some key assumptions here, which I examine further below. I do not explore this simple technique for metabolic rate determination further; the technique requires a lot of tedious (not to mention unpleasant) measurements and is now seldom used. Let us return to the candle and consider the two remaining calorimetric techniques.

INDIRECT CALORIMETRY

Indirect calorimetry measures the candle's heat production via the flame's O_2 consumption and CO_2 production. The technique is quite accurate, provided that good gas analyzers are available, and is the subject of most of this book. Indirect calorimetry is based on Hess's law of constant heat sums, which states that a chemical reaction (in this example, an exothermic or heat-producing reaction) releases a total amount of heat that depends only on the natures of the initial reactants and final products. Any intermediate steps are not relevant. Note that this assumption also underlies the energy balance example given above. For example, the full oxidation of glucose yields 3.74 Cal (or 3740 cal) or 15.6 kJ g^{-1} whether the oxidation (or catabolism) takes place in a yeast culture, a fruit fly, a mushroom, a shark, a petunia, a mouse, or a bomb calorimeter. Therefore, measuring metabolism in terms of heat production is possible if we know the respiratory substrate that was catabolized and the rate at which the end products (primarily O_2 and CO_2) were consumed or produced during the process. From these rates a thermal equivalent can be derived that yields heat production or metabolic rate.

For example, let us assume that we measure the oxygen consumption rate of the candle and find that it is 221 mL O_2 min^{-1}. This is a semi-legitimate and often-used unit for reporting metabolic rates. Suppose that we know further that 19.0 J of heat is released during the combustion of a particular type of wax by 1 mL O_2 at STP. In this case, the heat production rate of the candle is (221 $\times$ 19)/60 or 70 W. I explore this approach in far more detail in Chapters 7–14.

DIRECT CALORIMETRY

Direct calorimetry, measuring the heat produced by the flame directly, is the most obvious and accurate method. It is accurate not only because it relies on quite easily measured physical principles, but also because it includes all metabolism (anaerobic as well as aerobic) in its scope. In so doing, it makes a direct measurement of metabolic rate without any inferences based on gas exchange and metabolic stoichiometry. As such, and if carefully and properly applied (and there's the rub!), it can be regarded as the "gold standard" of metabolic measurement (Kaiyala and Ramsay, 2011).

Studies have shown that anaerobic contributions to the overall metabolism of some animals may be substantial (see Walsberg and Hoffman, 2005), so direct calorimetry will become an increasingly important research tool. This chapter discusses only a few of the most common calorimeter designs, chiefly applying to small mammals. I analyze gradient and differential calorimeters and the combining of direct and indirect calorimetry.

All approaches share some commonalities. As much as possible of an organism's heat production must be measured without regard to the animal's position in the calorimeter, so in the most common designs the organism is surrounded by a metallic container that distributes the heat evenly. However, a completely sealed calorimeter would soon kill the animal within it, so some air exchange with the outside world must occur. This air exchange can be used to make simultaneous respirometric measurements. However, we pay two penalties for allowing this. If we define the heat flux of the organism as Q, then the *measured* heat flux of the organism is:

$$Q_m = Q - Q_a - Q_w \qquad (6.1)$$

where Q_m is the measured heat production in watts, Q_a is the heat lost by heating the air entering the chamber, and Q_w is the heat lost by evaporation of water from the organism. Thus, these parameters will need to be measured and accounted for. They are covered in depth in the section "Gradient calorimetry."

Gradient calorimetry

The gradient calorimeter is the most common type of calorimeter. Its principle of operation is quite simple. A heat flux, Q, passes through a thermally conductive layer with a thickness D, area A, and specific thermal conductivity C. The result is a temperature difference, T, across the conductive layer that increases with D and decreases with C and A, thus:

$$T = QD/(CA) \qquad (6.2)$$

where T is in degrees Celsius, Q is in Watts, D is in meters, A is in square meters, and C is in Watts per meter-Kelvin [W $(m\ K)^{-1}$]. To get a feel for this, C for aluminum is about 237 W $(m\ K)^{-1}$, so if you transferred a watt of heat energy (from a rodent, say) through a piece of aluminum 1 cm thick and 10×10 cm in area, the resulting temperature gradient would be $(1 \times 0.01)/(237 \times 0.01)$, or $0.0042°C$. If you repeat this calculation with wood ($C \sim 0.1$), the gradient increases to $10°C$. Usually, gradient calorimeters use a thin layer of a fairly conductive substance and measure the resulting small temperature gradient with many (typically hundreds or even thousands) of thermocouples connected in series, referred to as a *thermopile*. Thermopiles obviously generate far larger (and thus easier to measure) signals than individual thermocouples:

$$V = nsT \tag{6.3}$$

where V is the voltage created by the thermopile, n is the number of thermocouple junctions connected in series, s is the Seebeck coefficient of the thermocouple junctions (for example, about 30 $\mu V°C^{-1}$ for copper/constantan junctions at $20°C$), and T is the mean temperature difference between the thermopile's two sets of junctions. One set of thermocouple junctions is on the inside and the other is on the outside of the gradient layer, and a box is built out of six gradient layers, one for each face of the box, the thermopiles of which are all connected in series (Figure 6.1). It is important that the six thermopiles cover the entire area of the calorimeter's six walls.

Even with many junctions connected in series, a thermopile's output is quite small, typically only a few millivolts to tens of millivolts in operation. It is usually a good idea to amplify a thermopile's output before connecting it to a data acquisition system. Suitable preamplifiers can be made easily by anyone with basic electronic knowledge (a circuit is shown in Figure 6.2), or they can be purchased.

An alternative to the classic thermopile is to use a Peltier effect heat pump, commonly used in 12-V DC-powered beer coolers, in reverse. These devices are thermopiles that

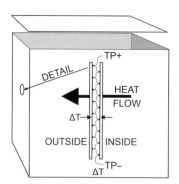

Figure 6.1 Simplified diagrammatic representation of a gradient layer calorimeter. Note that the thermopile (TP–, TP+) is shown for just a small area of one face of the calorimeter; typical thermopiles consist of hundreds to thousands of thermocouple pairs connected in series. An organism (not shown) in the calorimeter causes a heat flux through the calorimeter walls, resulting in a small temperature difference, ΔT, which generates a voltage across the thermopile by the Seebeck effect. A common construction technique is to connect the thermopile across a layer of insulating board held between two aluminum plates (e.g., Walsberg and Hoffman, 2005). The temperature of the outer layer of the calorimeter must be held constant. Air ports (not shown) are required if O_2 consumption over the period of measurement will be significant; they can also be used to enable indirect calorimetry in the same apparatus with some precautions (see text).

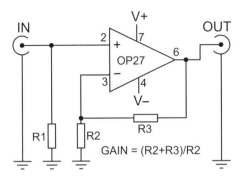

Figure 6.2 A simple circuit for amplifying the output of a gradient calorimeter thermopile. The gain of the circuit is set by the ratio of $(R_2 + R_3)$ to R_2. For example, if R_2 is 1 kΩ and R_3 is 100 kΩ, the gain is 101. Note that a split power supply must be used (a pair of 9-V batteries will work). R_1 is optional (typical value 10 kΩ) and merely serves to prevent the amplifier from "pinning" when the calorimeter is not connected. The operational amplifier type is not especially critical; the OP27 is a high-end choice made by Analog Devices and is available from Digi-Key (see Appendix 1).

consist of 63–127 antimony telluride thermocouples in series, and their Seebeck coefficients are staggeringly high. Unfortunately, they are quite small (typically 5 × 5 cm), making adequate coverage quite difficult.

Considering that the calorimeter is empty and you are recording data from it, you would expect to see a nice flat line close to zero. Instead you see a grotesque sinusoid with a period of half an hour or so. What can it be? Answer: The room's air conditioner is cycling, and with each cycle heat enters or leaves your calorimeter, creating a huge artifact signal. From this it is clear that the outer layer of a gradient calorimeter (outside the thermopile) must be kept at a constant temperature. This introduces a regrettable complexity to our design and is one of the reasons that gradient calorimeters are rare. Either the outer wall of the gradient layer calorimeter is equipped with wrap-around copper tubing attached to a water bath, or it can be directly immersed in a substantial bath of water. In either case the temperature of the water must be regulated extremely closely; even 0.01°C fluctuations in the water temperature can introduce unacceptably large fluctuations in the thermopile signal. Most owners of gradient layer calorimeters keep a large (200 L or so) water tank that feeds the calorimeter's water jackets and acts as a temperature buffer. Designing the six-sided, thermopiled, water-jacketed box is not easy, especially because the animal must be placed inside and afterward removed with a minimum of fuss. Typically, the top of the box is hinged and gasketed with closed-cell foam while maintaining good thermal contact on the outer edges. Input and output air ports are required, and care should be taken to ensure that they are positioned well away from each other if they might be used for indirect calorimetry (respirometry). The incurrent port should be supplied with a baffle or diffuser to assist in mixing the air in the chamber. In normal respirometry chambers, a small fan will help air mixing (see Chapter 8), but here a fan is problematic because it will generate heat and will also increase the thermal conductance from the animal (see Chapter 13). By all means try using a small fan if you wish, but validate its effect on the system carefully and make sure it is turned on while calibrating the calorimeter.

Differential calorimetry

The differential calorimeter measures heat flow between a chamber containing an animal and an adjacent unoccupied chamber of similar thermal mass (Figure 6.3). The two chambers are usually well insulated except on the single surface through which they exchange heat flow. A Peltier element operated "in reverse" (i.e., generating voltage from an applied heat flow) makes an excellent heat flow sensor. The two chambers are usually surrounded by an enclosure that is temperature controlled, for example, with a water-jacket.

Differential calorimeters can be made extremely sensitive, and the best are easily capable of measuring the metabolic heat production of brine-shrimp eggs (Hand and Gnaiger, 1988). On the other end of the scale, I have made a crude differential calorimeter out of a completely unmodified small thermoelectric beer cooler (Figure 6.4). Five of its six walls were insulated with polyurethane foam, while the sixth wall contained a heat sink attached to a Peltier heat pump, which in turn was connected to another heat sink on the outside of the cooler. In normal operation, an applied voltage pumped heat out of the cooler and dumped it into the external heat sink, from which it was removed by convection assisted by a small fan. Operated as a calorimeter, inlet and outlet air connections were added, the fan was disconnected, and the voltage produced by the Peltier heat pump was monitored. Calibration was performed with a resistor as described above, and a rat was introduced, which gave a satisfactory though not necessarily publication-grade signal even in the absence of external temperature stabilization. I mention this 3-h,

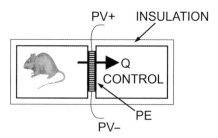

Figure 6.3 Diagrammatic representation of a differential calorimeter. In this example, a commercial Peltier effect thermoelectric heat pump module (PE; found in beer coolers, etc.) is used "in reverse" to generate a potential difference PV+ to PV− by the Seebeck effect as a portion of the heat flux, Q, from the organism passing through the module from the measurement to the control chamber. The chambers should be closely matched. Insulation is recommended, but care should be taken lest the animal become heat-stressed from its own heat production. Air ports are generally required but are not shown.

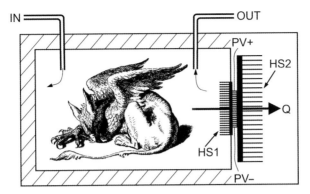

Figure 6.4 A crude differential calorimeter fabricated from a thermoelectric beer cooler with minimum modifications. The insulation and the two internal heat sinks (HS1 and HS2) are left in place, as is the original Peltier effect module, which is sandwiched between them. Air ports are shown. The voltage produced by the Peltier effect module, operating "in reverse" and generating the voltage between PE+ and PE– via the Seebeck effect, is proportional to the heat flux, Q, through the module. The system is easily calibrated using a resistor as in Figure 6.3. For good results, a stable ambient temperature is required.

proof-of-concept project only to emphasize that calorimeters are not necessarily mysterious masterpieces that take many months to create and fine tune, though they certainly can be in that class.

To further illustrate this point, consider a differential calorimeter, complete with flow-through indirect calorimetry, described by Lamprecht et al. (1998). They converted a plastic air-jacketed wine bottle holder into a water-jacket by adding entry and exit ports, cemented a Peltier effect heat pump to the bottom, and placed a can on top of the Peltier device. Using a pair of these contraptions outdoors, they were able to measure the heat production of Lotus flowers (one can contained a lotus flower, while the control chamber did not). They also used a more conventional gradient calorimeter in their study.

In another example, Jõgar et al. (2005) used a very simple differential calorimeter to examine gas exchange in pupae of the cabbage white butterfly. They constructed two 0.5-cm³ open chambers from 100-µm copper foil and joined them with a constantan wire, thus creating a single thermocouple. Using an extremely sensitive voltmeter, they measured the voltage between the two chambers, which they calibrated against heat production with a resistor. Using this device, they followed the respiratory patterns of a pupa (housed in one of the chambers) over many days and could track spiracular openings via the sudden heat loss events caused by water evaporation from the tracheal system.

Fluid transfer calorimetry

An alternative technique for measuring heat production is to enclose an animal in a thermally conductive layer surrounded by insulation and circulate a fluid between that layer and an external, thermally buffered reservoir. The heat production of the animal is then proportional to the temperature difference between the incurrent and excurrent fluid streams. This has the advantage that the temperature difference can be measured at a single location, rather than across the entirety of one or more (differential) or six (gradient) calorimeter walls. Buffering this advantage is the requirement for even conduction of the animal's heat production to the circulating fluid.

An example of this technique is described by Domenech et al. (1988). The authors constructed a 19×13×13 cm chamber from thin copper sheeting with a hinged lid. In a legerdemain feat of metalwork, they then soldered a wide square section tube also made of copper sheeting around all six walls of the chamber including the lid. Water at a controlled temperature flowed through this serpentine labyrinth with an interior volume of ~ 100 mL, and matched thermistors detected the increase in water temperature from the gravity-fed incurrent to excurrent flows (parenthetically, I would have recommended a thermopile and a peristaltic pump). Abundant thermal insulation was used throughout, and the animal was supplied with 0.3–1 L air min^{-1} from a tank source. The authors used a rather idiosyncratic calibration method involving sealed ~ 200-mL containers of water at various temperatures. With 0.001°C resolution of the incurrent–excurrent temperature differential, the authors obtained a calculated heat capture efficiency of 50 percent. Judging from their Figure 3 showing measured temperature differentials, the system gave reasonably clean data, at least for animals the size of rats. Sadly, the effective EE resolution cannot be calculated because the results were presented in joules per gram$^{-0.75}$ per minute but body masses were not reported (as is all too common in the biomedical literature; see Chapter 15!).

In a different implementation, Burger and van Breukelen (2013) used a coil of copper tubing surrounding the interior of an insulated canister. Rather than using water to transfer heat, they used mineral oil because its low specific heat (only 40 percent that of water) created a larger temperature differential for a given heat input. A large oil reservoir (20 L) minimized temperature fluctuations. To sense the temperature differential between the incurrent and excurrent oil streams they used a Peltier heat pump clamped between two CPU fluid-cooling heat sinks fed by a peristaltic pump at ~ 11 mL min^{-1}. Peltier modules are basically high-efficiency, approximately 127-element thermopiles and will happily operate in either Seebeck (heat flow sensing, generating a voltage) or Peltier (heat pumping, proportional to an applied voltage) mode. There is a minor design wrinkle; the authors found that enough heat was transferred from the excurrent to the incurrent stream through the Peltier module to distort the readings, so they used a separate incurrent shunt circuit. Their Peltier module generated about 3 V W^{-1} with good linearity across calibrations ranging from 0.25 W (mouse-like) to 3.25 W (rat-like), and the authors reported an effective EE resolution of ~ 0.09 W—adequate for larger rodents such as rats but marginal for mice. The authors incorporated indirect calorimetry into this apparatus but did not report any comparisons between their direct and indirect EE measurements (perhaps wisely; see the section "Comparing direct and indirect calorimetry").

CALIBRATING AND USING
A DIRECT CALORIMETER

Although making the calorimeter may be quite hard, calibrating it is relatively easy. All that is required is a known flow of heat. This is trivial to arrange:

$$Q = V^2/R \qquad (6.4)$$

where Q is heat flow in watts, and V is the voltage applied across a resistor R, where R is in ohms. For example, 5.00 V applied across a 10-Ω resistor dissipates $5^2/10$, or 2.50 W of heat. For best results, use a precision wire-wound resistor and a good-quality power supply. String the wires connected to the resistor through an air port (Figure 6.5). For best accuracy, measure the resistor's value using an accurate multimeter and allow for energy loss caused by wire resistance by making a four-wire measurement of the voltage across the resistor (Figure 6.3). You may want to use two or more resistors connected in series and/or parallel to spread out the heat source.

As the calorimeter warms up, the voltage produced by its thermopiles will rise to a plateau (Figure 6.6). At the plateau level, heat leaves the calorimeter through its walls at the same rate as it is being generated inside by the resistor. It may take a while to reach this point. This sluggish response time, typical of most direct calorimeters, can be substantially improved by application of the instantaneous transformation (Chapter 8; see also Figures 6.7 and 6.8 and accompanying discussion below). If

$$K = Q / V_t \qquad (6.5)$$

where Q is the calculated heat output of the calibration resistor in watts and V_t is the voltage produced by the calorimeter's thermopiles (after amplification, if applicable), then K, the calibration constant of the calorimeter, will have the units watts per volt. In

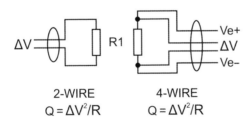

2-WIRE
$Q = \Delta V^2/R$

4-WIRE
$Q = \Delta V^2/R$

Figure 6.5 Using a resistor to calibrate a direct calorimeter by generating a known heat flux, Q, in watts as a result of applying a known voltage across it. Two-wire and four-wire connections are shown. In the two-wire connection, a voltage, ΔV, is applied across the resistor. Four-wire connections are more accurate because they eliminate the voltage drop in the leads going to the resistor. The voltage is now applied across excitation inputs Ve+ and Ve−. The two separate leads from the resistor to the measuring voltmeter carry negligible current and therefore give an accurate measurement of ΔV across the calibration resistor. A precise, stable resistor and an accurate digital voltmeter should be used.

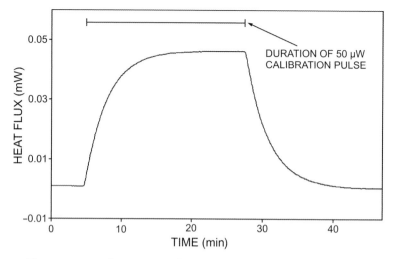

Figure 6.6 The raw output of a prototype ThermoMetrics differential calorimeter equipped with a 3.5-cm³ stainless-steel perfusion module when a 50-μW calibration heat pulse is applied (Lighton and Gnaiger, unpublished data). Note the long rise and fall times.

future, simply multiply the calorimeter's voltage output by K and you have the heat production of the animal within the calorimeter in watts.

This treatment neglects the heat loss caused by two factors that are important in actual calorimetric measurements, especially with larger animals such as mammals. These are convective heat loss, caused by the air flowing through the system, and evaporative heat loss, caused by the evaporation of water from the integument and the respiratory system of the animal (not to mention any urine or feces it may produce). I deal with each of these in turn.

To correct for evaporative heat loss (which is usually the larger source of error), you need to measure the water vapor pressures of the incurrent and excurrent air streams. Making the incurrent air as dry as possible (close to 0 percent RH) makes measuring the incurrent water vapor pressure unnecessary, but of course it maximizes any potential evaporative heat loss. You will now need to measure the rate at which the animal within the chamber is losing water. To a very good approximation,

$$V_{H_2O} = FR\,(F_eH_2O - F_iH_2O)/(1 - F_eH_2O) \tag{6.6}$$

where V_{H_2O} is the rate of water vapor loss in milliliters per minute, FR is the STP-corrected flow rate in milliliters per minute, and F_eH_2O and F_iH_2O are the excurrent and incurrent fractional concentrations of water vapor, respectively. These are calculated by dividing measured water vapor pressure (see Chapter 16) by ambient barometric pressure. This equation is a slight approximation because the flow rate is affected by the difference between an animal's consumption of O_2 (which reduces it) and production of

CO_2 (which increases it), which can usually be safely neglected in this case. If you want to include these terms, the equation becomes

$$V_{H_2O} = [F_eH_2O(FR + V_{CO_2} - V_{O_2}) - FR(F_iH_2O)]/(1 - F_eH_2O) \qquad (6.7)$$

where units are as before, and V_{CO_2} and V_{O_2} are the rates of CO_2 production and O_2 consumption of the animal, respectively (see Chapter 9 for the relevant equations).

Now, 1 mol of water weighs 18 g and in the vapor phase occupies 22.4 L, so the mass of water in grams that is lost by evaporation per second, V_w, is

$$V_w = [(18 \times V_{H_2O})/22{,}400]/60 \qquad (6.8)$$

or V_{H_2O} multiplied by 1.339×10^{-5}. The latent heat of vaporization of water is about 2.43 kJ g^{-1} at room temperature, so multiplying V_{H_2O} by $(1.339 \times 10^{-5})(2430)$ or 0.0325 yields heat loss due to evaporation in watts. Thus,

$$Q_w = V_{H_2O} \times 0.0325 \qquad (6.9)$$

where Q_w is the heat loss due to evaporation in watts (= joules per second). As an alternative, you can measure water loss rate in other ways and then use this line of reasoning to calculate heat lost by evaporation. For example, an animal might be losing 0.01 mL min^{-1} of water by evaporation. That doesn't sound like much, but it is equivalent to losing $(0.01 \times 2430)/60$, or 0.405 W, which may be a significant fraction of the animal's metabolic rate—indeed, more than the average EE of a normal mouse. The high efficiency of heat loss via evaporation (evaporative cooling) explains how a human, with an EE of 100 W and receiving a heat load of 500 W from sunlight at an ambient temperature of 40°C, can still maintain a core body temperature of 37°C; only evaporative heat loss by sweating makes this possible.

But adding to the methodological challenges of direct calorimetry, we're not only considering evaporation—there's also the heat transfer that accompanies urination or defecation. As Speakman (2014) states, "Half a milliliter of urine cooling down from 37 to 20°C over 10 min would spuriously increase the measured heat production by 36%."

Now we add a still further level of complexity: the storage and release of body heat. The body temperature of most small mammals is quite labile. As Speakman (2014) summarizes, a fairly typical 2°C body temperature swing for a mouse will store or release ~ 100 J of energy, enough to cause a substantial error in the direct calorimetry data of an animal with an EE of only ~ 0.2–0.3 J sec^{-1}. Compensating for this error is far from trivial, and at the very least requires measurement of body temperature, probably via telemetry, and reasonable assumptions about the specific heat of body tissues—which in turn requires knowledge of body composition.

I now deal with heat loss by convection. To measure this accurately, we need to know the temperature difference between the incurrent and excurrent air streams, and the heat capacity of the air. Obviously,

$$T_d = Q/[H(FR)] \qquad (6.10)$$

where T_d is the temperature difference between incurrent and excurrent air in degrees Celsius, Q is the heat dissipated in the calorimeter in watts, H is the heat capacity of air in joules per mole per Kelvin, which is about 29.2 in these units, and FR is the flow rate leaving the chamber in moles per second. Rearranging this,

$$Q_a = T_d H(\text{FR}) \tag{6.11}$$

where Q_a is the heat lost by convection. We can, for most purposes, assume that FR is equivalent to the flow entering the chamber. Thus, FR is STPFR(7.44×10^{-7}), were STPFR is STP-corrected flow rate in milliliters per minute. At 1 L min^{-1} flow rate, a temperature differential of 4°C corresponds to a heat loss of 0.042 W. This is a small but not insignificant figure.

We are making two simplifying assumptions here. FR is not the incurrent flow rate; it is actually the sum of the incurrent flow rate plus $V_{H_2}o$ and V_{CO_2} minus V_{O_2}. In addition, the heat capacity of air differs slightly with its composition; water vapor, for example, has about twice the heat capacity of dry air. The contribution of water vapor pressure to the total pressure of the air stream (and thus its flow rate) will seldom exceed about 1 percent at most, so the error introduced by our simplifying assumptions is negligible relative to other sources of error in the system, notably in flow rate measurement (see Chapter 17). $V_{H_2}o$ needs to be known for evaporative heat loss correction, which is not optional, and its heat capacity is about 37.5 J (mol K)$^{-1}$, so it is easily added to the convective heat loss budget if desired. For those interested in more information, a very thorough

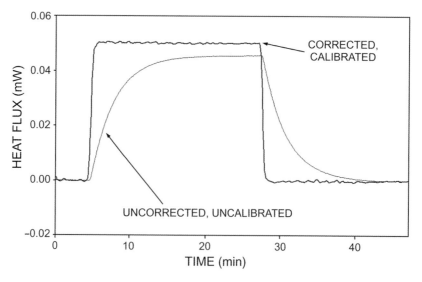

Figure 6.7 Applying the instantaneous transform or response time correction (thick trace) to the raw calorimeter data of Figure 6.6 (thin trace). The instantaneous-corrected data have also been scaled (spanned) so that the plateau level is equal to the calibration pulse value of 50 μW (0.05 mW). This correction is optimal and was iteratively fitted in ExpeData. If under- or overcorrection occurs, the instantaneous trace does not resemble the square-wave shape of the calibration heat pulse (see Lighton and Halsey, 2011, for discussion).

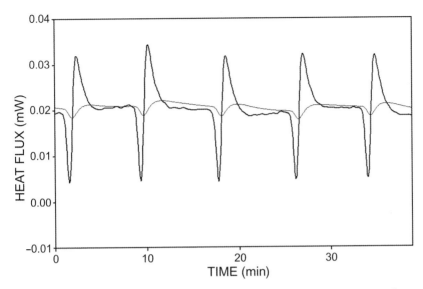

Figure 6.8 Signal processing using the Z transform helps reveal the dynamic nature of events in calorimetry. The raw data (thin trace) are grossly temporally distorted relative to the instantaneous-corrected data (thick trace). Rapid changes in heat flux are hardly visible in the uncorrected data. Note that the original and instantaneous-corrected data have exactly the same mean values over the recording. The correction coefficient used was calculated from a calibration pulse (e.g., Figure 6.7). Data were processed in ExpeData from an ant, *Camponotus vicinus* (Lighton and Gnaiger, unpublished data).

treatment of convective and evaporative heat loss in calorimetry can be found in Walsberg and Hoffman (2005).

For the vast majority of calorimetric projects, simply calibrating the calorimeter with air flowing through it at the chosen rate allows you to ignore the convective heat loss term, which can be assumed to be constant. The chief potential sources of error would be the higher heat capacity of water vapor given off by the animal and the addition of that water vapor volume to the incurrent flow rate. You now have the tools required to make these minor corrections if you wish.

Signal processing in calorimetry can be useful. As shown for the specimen calorimeter calibration curve (Figure 6.6), calorimeters can be quite slow-responding. However, adding the first derivative of a calorimeter recording (multiplied by a factor we refer to as Z) to that recording compensates for the response lag, as shown in Figure 6.7. This can easily be done during data analysis or even during data acquisition. Z is usually determined by introducing a known magnitude of heat pulse (e.g., from a calibration resistor), while recording the output of the calorimeter. During data analysis Z is iteratively changed to create a rectangular pulse with no over- or undershoots (Figure 6.7). An identical technique is used for improving response times in flow-through respirometry (Chapter 8). On occasion, signal processing is necessary to extract information on dynamic events. Figure 6.8 shows an example.

Comparing direct and indirect calorimetry

The bottom line is that if you understand the likely sources of errors, building a calorimeter is not necessarily terribly difficult; it can be done in multiple ways, and your ingenuity is the primary limiting factor. Using direct calorimetry, especially with endothermic animals like mammals or birds, ay, there's the rub, as you have seen above; and very especially when comparing direct and indirect calorimetry.

Do not even consider combining direct and indirect calorimetry with the intention of comparing them unless you are an ascended, ninth-Dan master or mistress of both techniques. Combining direct and indirect calorimetry is not a trivial task and I have only scratched the surface of the methodological issues that may confront you (see, for example, Speakman, 2013; Speakman, 2014).

As the careful work of Walsberg and Hoffman (2005) indicates, estimates of EE derived from indirect and direct calorimetry may, in some animals, suggest a significant anaerobic component of overall energy flux. The devil is in the details, however. To make such a comparison meaningful, both the direct and the indirect measurements must be as technically accurate as possible. This is an area of investigation where "sanity checks" on metabolic data derived from indirect calorimetry can be informative, as Figure 15.4 shows.

7

Measuring Field Metabolic Rates

ISOTOPIC TRACER TECHNIQUES

First described by Lifson et al. (1955), the doubly labeled water technique has proved to be an accurate and reliable method for measuring the metabolic rates of unrestrained mammals and birds. It relies, ingeniously enough, on the differing fates of hydrogen and oxygen in water consumed by (or injected into) an organism. Hydrogen stays almost exclusively associated with water alone, and the amount of hydrogen originally ingested slowly declines over time as water is lost through evaporation or excretion and the original water is replaced. The decline follows classic first-order rate kinetics, where a constant fraction of the total water content of the body is lost per unit time, following the curve proportional to e^{-kt}, where k is the rate constant and t is elapsed time.

The situation is different for oxygen, which mingles with the oxygen in CO_2 as well as the oxygen in H_2O. This equilibrium is catalyzed by carbonic anhydrase in the red blood cells and the lung, which shuffles oxygen randomly between H_2O and CO_2 via the bicarbonate anion. Consequently, the oxygen in a given bolus of H_2O is lost more rapidly than the hydrogen because some of it is breathed out as the oxygen in CO_2. The kinetics of oxygen and hydrogen are both first-order, but the rate constant is higher for oxygen. The degree of increase allows us to estimate the animal's rate of CO_2 production and thus, more or less, its metabolic rate.

In modern investigations, hydrogen in the tracer bolus is replaced by deuterium (^2H) and ^{16}O is replaced by ^{18}O. These isotopes can easily be distinguished from their more common siblings by mass spectrometry, thus giving the owners of these instruments a great advantage. In a typical doubly labeled water study, a baseline sample of blood is withdrawn from the animal, and the bolus of rather expensive labeled water is injected. Another blood sample is taken after half an hour or so to fix the initial concentrations of the labels. The animal is then released and recaptured after a day or two, and a third blood sample taken. The three samples are sent to the mass spectrometry lab, and the concentrations in parts per million of ^2H and ^{18}O in the three samples are returned. The calculations that ensue are quite simple.

$$VCO_2 = \left[(N/2.08)(K_o - K_d) \right] - \left[(0.015)(K_d)(N) \right] \tag{7.1}$$

Measuring Metabolic Rates: A Manual for Scientists. Second Edition. John R. B. Lighton, Oxford University Press (2019).
© John R. B. Lighton (2019). DOI: 10.1093/oso/9780198830399.001.0001

where

$$K_o = \left[\ln\left(I_o - B_o\right) - \ln\left(F_o - B_o\right)\right] / t \tag{7.2}$$

and

$$K_d = \left[\ln\left(I_d - B_d\right) - \ln\left(F_d - B_d\right)\right] / t \tag{7.3}$$

and where V_{CO_2} is the animal's rate of CO_2 production in millimoles per hour, N is the animal's total body water content in millimoles, K_o is the fractional turnover of oxygen per hour, K_d is the fractional turnover of hydrogen in the same units, I_o is the initial concentration of ^{18}O, F_o is the final concentration of ^{18}O, and B_o is the background concentration of ^{18}O, all in parts per million. I_d, F_d, and B_d are the analogous figures for deuterium. Finally, t is the time between the initial and final samples in hours. The equation takes fractionation effects, caused by the different behaviors of the isotopes from their more common siblings, into account. Details can be found in Speakman and Racey (1986), Speakman (1998), and Blanc et al. (2000).

Note that you need to know N, the total body water content of the animal. It is divided by two (plus a fractionation fudge factor) to reflect the fact that a mole of CO_2 removes two atoms of oxygen to just one via a mole of H_2O. It is usually estimated from the change in ^{18}O concentration between the first (preinjection) and second (shortly after injection) blood samples. Thus,

$$N = 55.5\left(VM / D_o\right) \tag{7.4}$$

where N is the body water content in millimoles, V is the volume of the injection in milliliters, M is the ^{18}O concentration of the injection, and D_o is the difference in pre- and postinjection ^{18}O concentrations. This is a simplification because ^{18}O can participate in other unmodeled activities, but it does so to a smaller extent than ^2H, so this equation is accurate to about 2 percent (Mata et al., 2006). The amount of doubly labeled water should be tuned to the size of the organism, making the use of small organisms attractive, especially in the early days of the technique when enough doubly labeled water for a 70-kg mammal such as a human would have cost $50,000, not adjusted for inflation. Doubly labeled water is now inexpensive enough to be a standard method of investigating human obesity (Speakman, 1998).

In general, doubly labeled water gives accurate results when compared with measurements made via indirect calorimetry. Nagy (1989) has composed an impressive list of statistics in this area. The major caution when using doubly labeled water is that CO_2 production, not O_2 consumption, is measured. This means that the energy equivalent (the real metabolic rate!) corresponding to a given V_{CO_2} can vary > 30 percent or so, depending on the dominant respiratory substrate that the animal uses.

FLOW-THROUGH RESPIROMETRY

Provided that an animal can be persuaded to enter an area from which air can be pumped at a known flow rate and sampled for gas analysis, you can, in principle, measure the animal's metabolic rate with good precision without its knowledge. Basically, you are engaging in pull respirometry with a mask (see Chapter 11). The mask can be of your own fabrication, such as an enclosure, feeder, or nesting box, or it can be made by the animal itself, such as a burrow. As described in Chapter 11, because you are pulling air from a mask, you are forced to measure flow rates downstream from the animal, so your flow measurements will be influenced by the water vapor added to the air stream by the animal and by the O_2 consumed and the CO_2 produced by the animal. Likewise, O_2 will be diluted by water vapor and CO_2, and CO_2 will be diluted by water vapor and enriched by the extraction of O_2 consumed by the animal. The approaches to compensating for these effects are described in Chapter 11.

That said, noninvasive field-based respirometry is an important and entertaining technique. It avoids the stress responses that come with capture and restraint and allows the animal to behave naturally while measurements are made. I describe three studies that have used noninvasive respirometry and urge you to try it out when the opportunity presents itself.

The first study involved hummingbirds. Previous investigation of hovering energetics in hummingbirds had used chambers within which the birds hovered. Being, by and large, constant volume studies (Chapter 4), they also required careful timing of the actual hovering durations, plus an assumption of nonhovering metabolic rates (e.g., Lasiewski, 1963). Even where feeders were modified to function as masks (e.g., Epting, 1980), tunnel and ground effects associated with powerful movements of air within the confines of a respirometer chamber were inevitable, as were problems associated with constraining the flight movements of the birds, unnatural durations and patterns of hovering, and the condition of the captive birds.

By designing a feeder from which air could be pulled and which effectively enclosed the bird's nares while hovering (Figure 7.1), Bartholomew and Lighton (1986) measured

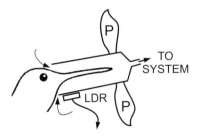

Figure 7.1 Diagram of a hummingbird feeder serving as a pull-mode respirometry mask. A plastic tube such as a pill vial is decorated with red petals (P) and equipped with an exit port from which air is pulled, typically at 1–2 L min^{-1}. A light-dependent resistor (LDR) allows the presence or absence of the bird's head in the feeder to be recorded (see Bartholomew and Lighton, 1986, for details).

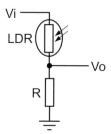

Figure 7.2 Using a photoresistor as an activity/position detector. Photoresistors, also known as light-dependent resistors (LDRs), are available from electronics distributors (see Appendix 1). The value of resistor R should be approximately equal to the mean of the LDR's range of resistances in the anticipated application. The input voltage, V_i, should not exceed the maximum voltage measurable by the data acquisition system. The resulting voltage, V_o, can be recorded synchronously with other data. Phototransistors or reverse-biased photodiodes can be used in place of the LDR for faster response times (see Horowitz and Hill, 2015).

the metabolic cost of hovering in free-flying wild Anna hummingbirds. The feeder/mask was suspended just outside the study of Bartholomew's house in the Santa Monica foothills; the gas analysis equipment was set up in the study. We pulled air through the system at 1 L min^{-1}. I wrote a program that monitored the O_2 concentration of the air stream and stored the results in a circular buffer. When the program detected a sudden decrease in O_2 concentration, it would continue sampling for several minutes and then save the data to disk, including a baseline period of 14 sec before the O_2 decrease. This meant that we could sip chilled dry sherry while the computer did the work. A photoresistor (Figure 7.2) provided a simultaneous record of the presence or absence of the bird's head in the mask. This was necessary because the birds generally interrupted their feeding bouts several times. This resulted in several distinct peaks of O_2 consumption which, when integrated to yield total volume of O_2 consumed and divided by total mask time, yielded hovering metabolic rate. This technique is now in wide use, especially since the compact, portable gas analysis systems now available have made it far easier to deploy (e.g., Welch and Suarez, 2007; see Groom et al., 2018, for a comprehensive review).

The second example involves animal behavior with which we're all familiar—column foraging in ants. Although it's possible to measure the energetics of running in ants with a tiny treadmill (Lighton et al., 1987) or a running-tube (Lighton and Feener, 1989), the conditions of measurement differ fundamentally from an ant locomoting voluntarily across a natural substrate. Because most ant species forage long distances from a central location, the energy cost of foraging is an important aspect of their biology. Accordingly, I devised a respirometer mask through which the ants could forage (Figure 7.3). The mask was of a known length, and while walking through it, the foraging column's CO_2 emission rate could be measured using a pull system housed beneath the shade of a nearby creosote bush. Baselines were provided by an adjacent, empty chamber of the same size. Video analysis allowed the number of ants walking through the chamber to be

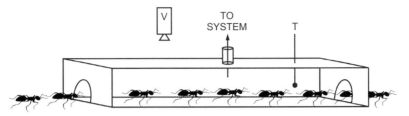

Figure 7.3 A simplified representation of a field system for measuring ant trail energetics. The ants follow a foraging trail through a respirometer chamber from which air is pulled at 1 L min⁻¹. A video camera (V) allows each ant's walking speed and body length (from which body mass can be calculated) to be determined. A thermocouple (T) allows ambient temperature to be measured. V_{CO_2} of the ant trail is directly measured and can also be calculated on the basis of the other measured parameters using separately acquired laboratory data. For more details, see Lighton and Duncan (2002).

counted, their walking speeds measured, and their body masses determined (via allometric analysis of the relation between body length and mass in that species). In combination with temperature recordings, it was therefore possible to use the body mass, walking speed, and ant count data to estimate what the metabolic rate of the foraging column section should be on the basis of extrapolation from laboratory-measured cost of transport, metabolic rate, and temperature sensitivity data. This figure could then be calculated from laboratory data over a succession of different column densities, speeds, and temperatures and compared against the actual measurements made on the column (Lighton and Duncan, 2002). It turned out that laboratory measurements accurately predicted field energetics. Stochastically and serendipitously, this system also provided the first-ever quantitative data on an ant colony's response to an earthquake, which was "shaken, not stirred" (Lighton and Duncan, 2005).

This example is cited because much that occurs in field respirometry does not make it into the literature, and it is useful for would-be field respirometrists to be prepared for the unexpected. When deployed, the system at first gave unusably unstable results, which we quickly discovered was caused by CO_2 seeping from the bare ground at rates that dwarfed the CO_2 output of the foraging column. This was solved by forcing the ants to forage across a sheet of Masonite, the rough side of which was similar in texture to the ground they traversed naturally. We quickly discovered that the ants, though easily visible when the video was replayed at normal speed, "disappeared" in freeze-frame. Matte white paint on the Masonite solved that problem. Next, we discovered that even a very slight breeze would cause a noticeable drop in the measured column metabolic rate, even though column activity was constant. The breeze was blowing the CO_2-enriched air out of the chamber. This necessitated a trip to the nearest hardware store for parts to make a PVC tubing frame and thin plastic glazing with which to cover it. This eliminated the effects of wind. Then, on the exact morning when everything was working to our satisfaction, the Landers earthquake occurred (magnitude 7.4 on the Richter scale), and we were able to measure, for the first time, the effect of a major earthquake on ant activity and metabolism—which turned out to be nonexistent! This created an interesting quandary. Though the earthquake aspect of this study was eventually published (Lighton and Duncan, 2005), doing so was not at all easy because of the negative nature of the result. One well-known ant biologist from Stanford gave three reasons for vehemently rejecting an earlier version. First, the paper lacked an hypothesis regarding the effect of earthquakes on ants that had been formulated in advance of the measurements. Second, the number of earthquakes, being only one, was inadequate. And third, the findings of the paper contradicted the folk wisdom of historically marginalized peoples, and were therefore highly suspect.

The last example involved the mud-dauber wasp (*Sceliphron caementarium*). Mud daubers build nests of mud that they provision with paralyzed spiders, which provide food for the mud dauber's larvae. The wasps collect balls of mud from the bank of a body of water, fly to the nest, form the mud (using their flight motor to vibrate the mud so that it flows easily), and repeat these operations until the nest is complete. To measure the energetics of nest building, including the vibrating mud forming, we constructed a Lucite mask with a doorway just large enough for the wasp to enter. We waited until a wasp had started nest construction, then placed the chamber over the nest while she went to collect the second ball of mud. On her return, the wasp walked around the chamber, finally releasing her mud ball, which had dried to the point where it was unusable for its intended purpose. Eventually the wasp discovered the doorway, entered, checked the nest, then left the chamber and flew away to obtain a brand-new mud ball. From that point on she returned to the chamber with successive mud balls and unerringly entered the doorway with no delay. This behavior held true for several wasps and nests. Figure 7.4 shows typical results.

Baselines, analogously to the hummingbird example, were automatically provided by the absence of the wasp from the chamber when she left to fetch another mud ball. As with the ant example, only CO_2 was measured. However, the measurements took place indoors, unlike the case with the ants. Consequently, as onlookers and experimentalists came and went, the baselines shifted. Figure 7.5 shows how interactive interpolation can correct shifting baselines. This example is included to emphasize the importance of flexible baseline correction in high-sensitivity field respirometry. It doesn't matter whether the shifting baselines are actual or instrumental in origin; if the system is baselined often, even quite serious drift can be corrected. If the baselining is intrinsic (occurring when the animal leaves the system), so much the better.

In summary, field respirometry is not only possible, but also quite easy under the right circumstances. With the advent of easily portable gas analyzers, data acquisition systems, and laptop computers, the chief limitation is the ingenuity of the researcher and the researcher's knowledge of the tradeoffs involved in pull-mode respirometry (Chapter 11). If you are investigating a suitable biological system, you should seriously consider moving your laboratory into the field, or at the very least supplementing your lab work with field data.

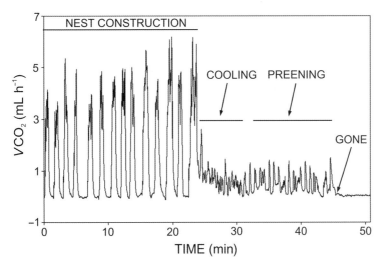

Figure 7.4 Nest construction by a mud-dauber wasp, *Sceliphron caementarium*, at Zzyzx, California. The wasp constructed its nest within a pull-mode respirometry mask placed over the nest when construction began (Lighton and Duncan, unpublished data). In the first half of the recording, the wasp repeatedly entered the chamber with a mud ball and added it to her nest, then left to collect another mud ball. During each absence the system returned to baseline. After nest construction was complete, her MR dropped sharply. She remained in the chamber for a few minutes while preening herself (and engaging in discontinuous gas exchange cycles; Lighton, 1996), then left, reestablishing the baseline.

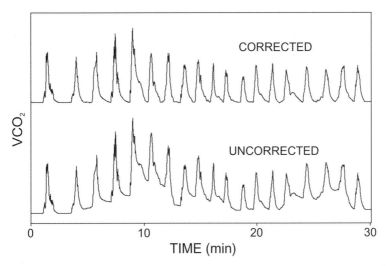

Figure 7.5 Correction of pathologically drifting baselines caused by people entering and leaving the recording area, in another recording using the same setup as in Figure 7.4. Interactive shifting baseline correction was used (ExpeData software). Such extreme baseline correction needs to be done cautiously, but any errors introduced are unlikely to be systematic.

This technique is by no means limited to the field. It could also be a valuable adjunct to biomedical research projects that require EE measurements of individual animals in a communally housed setting. Advantage can be taken of the animal's exploratory behavior to briefly isolate it in a subset of the system from which air is pulled at a high flow rate. In that case, readings from the unoccupied chamber before and after occupancy will serve as the baseline. Excellent gas analyzer precision will be required to maintain a small but accurately measurable ΔO_2 and/or ΔCO_2 at an acceptably high pull flow rate.

8

Flow-through Respirometry: The Basics

Flow-through respirometry has advantages and disadvantages, though the former out-weighs the latter by a wide margin in most applications. The primary disadvantage of flow-through respirometry is the sensitivity and stability (and thus cost) of the gas analyzers required to make it work reliably. This is not an intrinsically low-tech and postapocalyptic technology, unlike the simple manometric pressure-based techniques. Flow-through respirometry also requires the investigator to analyze the system's config-uration and make intelligent decisions about the contributions of various sources of error, though, to be fair, it's possible to adopt a cookbook approach for subsequent users of a system that is used in a stable configuration. Against these disadvantages is the fact that flow-through respirometry yields dynamic information on variations in metabolic rate. This allows, for example, the metabolic cost of a given activity such as running at a specific speed to be determined and for inactive metabolic rate to be teased apart from active metabolism in an animal that may alternate frequently between these states. When measuring both O_2 and CO_2, it's also possible to track dynamic changes in respi-ratory substrate use, such as, for example, the shift from fat to carbohydrate catabolism that occurs within minutes of a hummingbird's first nectar meal after a night of fasting (Suarez et al., 1990). A review article (Lighton and Halsey, 2011) covers these topics in a briefer way than will be achieved here and is a good first introduction to the topic.

All flow-through systems derive their information from the changes an organism cre-ates in the composition of an air stream, as measured by gas analyzers. To provide accu-rate data, the flow rate must be known, the gas analyzers must be accurate and recently calibrated, and the effects of the system's configuration must be known. I cover gas analyz-ers in Chapter 16. The principles and practice of flow-rate measurement are covered in Chapter 17. Basically, the major configurational differences between various flow-through respirometry techniques revolve around whether flow is pushed or pulled, where flow is measured, whether scrubbers for water vapor or CO_2 are used, and if so, where they are used. I will now examine the basic configurations. After this broad overview, I conclude the chapter with a checklist to complete before starting a new flow-through respirome-try project. Chapter 9 is devoted to respirometry equations, Chapters 10 and 11 cover the common variations of flow-through respirometry, Chapter 12 covers human metabolic rate measurement with an emphasis on room calorimetry, Chapter 13 covers metabolic

Measuring Metabolic Rates: A Manual for Scientists. Second Edition. John R. B. Lighton, Oxford University Press (2019).
© John R. B. Lighton (2019). DOI: 10.1093/oso/9780198830399.001.0001

phenotyping of common model animals such as mice and rats, and Chapter 14 discusses validation techniques.

PUSH MODE VERSUS PULL MODE RESPIROMETRY

There are two ways to move air past an animal—by pushing or pulling (see Figure 4.7). In either case, depending on the particular setup, we can measure the flow rate upstream or downstream from the animal. The exception to this rule is the mask (and, in practice, almost all field respirometry; Chapter 7), which, being open to the air, allows only the downstream measurement of flow rates pulled from it.

Push systems require a sealed respirometer chamber. Flow rates can be measured before or after the sealed respirometry chamber. *Pull systems* are often easier to set up than push systems. If flow in a pull system is measured upstream from the respirometry chamber, it is especially important to keep leaks to a minimum; thus, flow is usually measured downstream from the respirometry chamber or mask in pull systems. Likewise, measuring small animals against a zero-CO_2 baseline (typically created by scrubbing CO_2 from the incurrent air) requires excellent system integrity in a pull system, which is immersed in CO_2 at ambient intrabuilding concentrations of 500–1000 ppm.

Whether you are running a push or a pull system, you will need to measure flow rate, and the position of flow rate measurement is the primary determinant of which respirometry equations are appropriate. Measurement of flow rate is treated in detail in Chapter 17. Systems in which flow rate measurement takes place upstream from the respirometry chamber (in the incurrent air stream before the animal) are covered in Chapter 10; those in which flow rate measurement takes place in the excurrent air stream are covered in Chapter 11. Mask systems (a variation of pull respirometry with compulsory downstream flow measurement) are also covered in Chapters 11 and 12.

Whichever approach is used, there are certain commonalities. These include the selection of an appropriate flow rate, which is an exercise in compromise. Too low a flow rate and you run the risk of suffocating the animal or at least shifting its physiological state off the normal curve. In addition, a very low flow rate will increase the time required for the system to respond to changes in the animal's metabolism. Too high a flow rate may reduce the gas concentration signals below the threshold of usability because all gas analyzers drift to some extent. If analyzer drift or noise is comparable in magnitude to the signal from your animal, you are in trouble. This is less of a problem with CO_2 than with O_2 analyzers because the latter are tasked with measuring tiny changes from a large signal, whereas the former measure changes that are often much larger than background levels.

FLOW RATE: A GAS ANALYSIS PERSPECTIVE

Larger and more metabolically intense animals (such as mammals) will require higher flow rates than smaller animals with lower metabolic rates. As a rough guide, a reasonable flow rate should cause an O_2 depletion or CO_2 enrichment of between 0.01 percent (at the low side) and 1 percent (at the high side). The effect of a 1 percent depletion of O_2 will be negligible in most organisms, but this does not necessarily hold for a 1 percent increase in CO_2. Some organisms are capable of tolerating higher CO_2 levels than others; some mammals, for example, may increase ventilation frequency significantly at CO_2 levels approaching 1 percent. To be on the safe side, flow rates should be generated to yield typical O_2 depletions of around 0.1–0.5 percent at most. This minimizes the effect of analyzer drift on the low depletion side and of hypoxia and hypercapnia on the high depletion side. A calculator for suggesting flow rates, depending on the taxon and body mass of the experimental organism, is available at www.respirometry.org.

We can also take a more rigorous approach. The first step in selecting a flow rate is to consider the animal mass and taxon, which can be used to predict the animal's VO_2. Numerous equations exist for this purpose; a collection of several common ones is incorporated into this book's companion website, www.respirometry.org. The approach is simple:

Calculate the predicted VO_2 of your experimental animal.

Choose a standard respirometry equation (e.g., $VO_2 = FR(\Delta O_2)/(1 - 0.2094)$, where FR is flow rate and ΔO_2 is the difference in fractional concentration between incurrent and excurrent O_2).

Choose a reasonable value for ΔO_2, suitable for easy measurement using your equipment, such as 0.1 percent $= 0.001$ expressed as a fractional concentration. For an excellent O_2 analyzer, especially with frequent baselining, this figure can be reduced tenfold.

The recommended flow rate is $(VO_2)(1 - 0.2094)/\Delta O_2$.

For practical purposes, especially if you are also recording CO_2, you can simplify this to $VO_2/\Delta O_2$.

As an example, let us assume that you are working with a small mammal such as a mouse at standard room temperature with a predicted VO_2 of 1 mL min^{-1}. To obtain a predicted VO_2, you can use the collection of metabolic calculators available on the web site www.respirometry.org, choosing the one most suited to your experimental organism. Let us say that you feel comfortable, based on your experience with your respirometry system, with operating at an O_2 deletion of 0.05 percent. In that case, the recommended maximum flow rate would be approximately $1/0.0005$ or 2000 mL min^{-1}. Depending on the quality of your respirometry system and your degree of facility in using it, you might choose anything from a minimum of 70 mL min^{-1}, at which point hypercapnia will likely adversely affect the animal, if for whatever reason you are recording with cheap equipment with a high noise to signal ratio, to upwards of 10 L min^{-1} if you require high temporal resolution, own a state-of-the-art, low-noise system, *and* know how to use it— which mostly consists of scrupulous baselining.

Why the emphasis on baselining? The higher the flow rate and the closer you approach the noise and drift floor of the gas analyzers, the more critical baselining becomes. "Baselining" simply means passing incurrent, rather than excurrent, air through the system at intervals. This establishes the value from which the excurrent gas concentration values depart, allowing analyzer drift to be compensated for and, in most cases, essentially eliminated if suitable software is used. Techniques for baselining are covered in Chapters 10 and 11.

FLOW RATE: A RESPONSE TIME PERSPECTIVE

A convenient measure of the response time of a system can be obtained by dividing chamber volume by the flow rate of air through the chamber. For example, dividing a chamber volume of 1000 mL by a flow rate of 1000 mL min^{-1} yields a time-constant of 1 min. This is the time required for the system to respond to 63 percent (or $1 - e^{-1}$ expressed as a percentage) of a step change. In two time-constants (2 min in this case), the step change has reached 86 percent of its final value; after three time-constants it has reached 95 percent of its final value. In more general terms, the time, T, in minutes required to reach F% of the final value is:

$$T = -(V/FR)\ln\left[(100 - F)/100\right] \tag{8.1}$$

where V is the volume of the chamber in milliliters, FR is the flow rate in milliliters per minute, and ln denotes the natural logarithm. If you use base 10 logarithms, multiply the result by 2.302 (the natural log of 10).

It is therefore advisable to keep flow rates high enough, for a given chamber volume, to track changes in the animal's metabolism with the time resolution you require. If you are interested in tracking very fast-changing phenomena (e.g., in an energetics investigation), you will probably need to ramp up flow rates as high as practical, given the tradeoffs outlined above, or use the "instantaneous transform" (see "The instantaneous (or Z) transform" section of this chapter). If you are only interested in a measure of routine metabolic rate, then a much longer time constant will suffice.

AVOIDING POOR MIXING

The native time-constant of a system is not the whole story. Air is susceptible to boundary layer effects and may show poor mixing if it flows too slowly through a chamber. Under these circumstances the orderly and simple washout kinetics break down and become chaotic. Tendrils of deoxygenated air snake about the chamber and induce sudden spikes when pulled through the analyzers, dismaying the hapless investigator. As a general rule, keeping the time-constant of a system to a minute or less helps guard against the problem of poor mixing.

Perhaps the simplest method of ensuring adequate mixing at moderate to long time-constants is to arrange the incurrent and excurrent flow paths so that they are thoroughly separated and deliberately misaligned. For example, air might enter a chamber

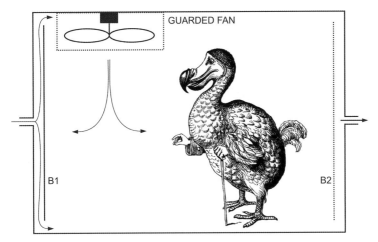

Figure 8.1 Using a baffle system in a respirometer chamber to encourage turbulent flow, which improves mixing characteristics and reduces boundary layer effects. A small fan (which should be surrounded by a perforated guard) also improves mixing, but its possible effects on thermoregulatory energy expenditure should be considered (see text).

and be spread out by a baffle or other obstruction, rather than flowing straight toward the outlet (Figure 8.1).

If a straight flow path between inlet and outlet is available, extraordinarily bad mixing artifacts are guaranteed. Diffusers, such as mufflers or airstones, can also help at low flow rates where their resistance to flow is insignificant. If your combination of chamber and flow rate yields a longer time-constant than about 1 min, allowing the excurrent air to emerge from a distributed manifold with many intake points provides a close equivalent to convective mixing without requiring mechanical air movement. If the time-constant of a chamber is more substantial and it is not practical to improve mixing in any other way, many investigators favor placing a small electric fan (such as the 12-V DC-powered fans sold for cooling computer CPUs and power supplies) in the chamber, guarded, if necessary, with some perforated metal to protect the fan or the chamber inhabitant from sustaining serious damage in case of contact. Vendors of such fans can be found in Appendix 1. The smallest, quietest fan that will provide adequate mixing should be used. Be aware that if a mammal or bird (i.e. a homeothermic endotherm) is in that chamber, the fan will increase convective heat loss from their body. Because they are defending a fixed body temperature, this will necessarily raise their metabolic rate to compensate for the additional heat loss. See Chapter 13 for a cautionary note on using convective air mixing in animal cages. It is amusing though depressing to note that such fans are a standard feature of some metabolic phenotyping systems, which often compound the thermoregulatory problems caused by forced convective air movement by housing experimental animals at a temperature at which they are cold-stressed (see Chapters 13 and 15).

THE INSTANTANEOUS (OR Z) TRANSFORM

Because of the simple first-order kinetics underlying washout phenomena, it is easy to increase the response speed of most respirometry systems by simple mathematical processing if speed is considered important. There is a caveat: For this approach to work acceptably well, good mixing of the air in the chamber is not optional; it is essential. Any chamber with a time-constant of less than about a minute is suitable; alternatively, mixing can be enhanced mechanically as described above, or by use of widely distributed sampling locations, such as via a circumferential tube with numerous microperforations from which excurrent air is pulled.

The technique, as described by Bartholomew et al. (1981; see also Lighton and Halsey, 2011, and Lighton, 2012), is often referred to as the "instantaneous correction," though it is far from instantaneous. The technique depends on the fact that for a given chamber volume and rate of air flow, the fractional rate at which equilibrium is approached remains constant regardless of the distance of the system from equilibrium. In other words, during a given interval of time, the system will change by a constant fraction of the distance to the new equilibrium value. Defining Z as that constant fraction,

$$Z = 1 - e^{-FR(dT/EV)} \qquad (8.2)$$

where FR is flow rate, dT is the interval between successive samples, and EV is the "effective volume" of the system, which will vary somewhat from its physical volume depending on the mixing and washout characteristics of the system, plus the response times of the analyzers and the volumes of any scrubbing columns participating in the event. Likewise,

$$Z = (F_e T - F_e T_{-1}) / (F_{eq} - F_e T_{-1}) \qquad (8.3)$$

where $F_e T$ is the fractional excurrent concentration of O_2 at time T, $F_e T_{-1}$ is the same concentration at time T_{-1} (one sampling interval previous), and F_{eq} is the final equilibrium excurrent O_2 concentration. Combining the two equations,

$$F_{eq} = [(F_e T - F_e T_{-1}) / (1 - e^{-FR(dT/EV)})] + F_e T_{-1} \qquad (8.4)$$

This equation looks rather fancy but is actually very simple to implement. Taking out the complicated terms and replacing them with a constant, we find that

$$F_{eq} = K(F_e T - F_e T_{-1}) + F_e T_{-1} \qquad (8.5)$$

In other words, (1) take the first derivative of the O_2 data, (2) multiply the first derivative by a constant, and (3) add the result to the original data. Voilà, the response time is greatly improved (Figure 8.2). It remains only to determine the value of the constant.

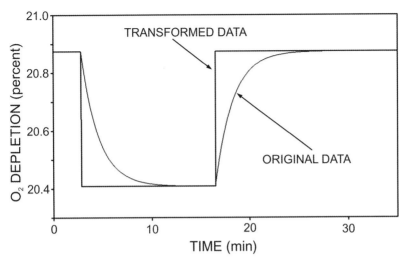

Figure 8.2 The effect of the instantaneous transformation on the response characteristics of an O_2-based respirometry system. The data before transformation simulates the response of a respirometry system with a time constant of 2 min (e.g., a 2-L chamber and flow rate of 1 L min^{-1}) to a hypothetical instantaneous change of an organism's Vo_2. The output of an ideal (noiseless) O_2 analyzer was simulated in ExpeData and then subjected to instantaneous transformation (see Chapter 8).

This is best done by iteration on a recording of O_2 concentration over time, following a sudden square wave disturbance in the O_2 concentration in the chamber, for example by injecting a small volume of nitrogen into it. The value of K is iteratively changed until a clean response is achieved with no under- or overcompensation.

Because this technique uses the first derivative of the O_2 reading and because many O_2 analyzers are quite noisy, you will need to take special care to get the cleanest possible signal from your gas analyzer if you use this technique. Good-quality analyzers allow variable filtration of their signals, and if a filtration value is chosen that is smaller than the analyzer's native response time (e.g., a filtration time-constant of ~ 2 sec for an analyzer with a native response [10–90 percent of a step change] of 10 sec), the filtration will reduce noise without significantly affecting response time. Likewise, good-quality analyzers have suppressed-zero analog output ranges which, for example, fit the range of 19–21.5 percent O_2 into a 5-V span. This can result in a fortyfold increase in effective resolution over the more common range of 0–100 percent over a 5-V span because the effect of quantization noise during data acquisition is proportionately reduced. In addition, good data acquisition systems offer the option of oversampling one or more channels or, in other words, taking numerous, very rapid samples from an instrument and then storing only the average. This reduces random electrical noise by a factor proportional to the square root of the number of readings. Thus, for example, taking 100 readings very rapidly (which only takes a tiny fraction of a second) can reduce random noise by a factor of 10 without any effect on response speed because the averaging takes place strictly within each sample. Finally, it is also possible to smooth data during analysis before taking

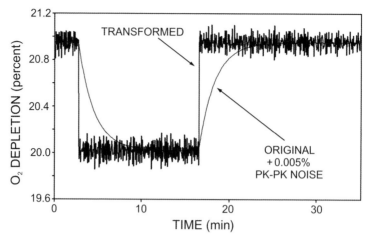

Figure 8.3 The effect of analyzer noise on the instantaneous transformation. Data are as in Figure 8.2, but before the instantaneous transformation, 0.005 percent peak-to-peak random noise was mathematically added to the signal to simulate a poor-quality O_2 analyzer and data acquisition system. This noise is hardly visible for the untransformed data but severely affects the transformed data. Noise can be reduced by smoothing before transformation at the expense of response time. Little if any smoothing is required when processing the output of a high-quality O_2 analyzer.

its first derivative, and in fact this is usually a good course of action. Figure 8.3 show an example of applying the instantaneous transform on a noisy signal similar to that produced by a poor-quality O_2 analyzer.

It is worth mentioning that judicious application of the instantaneous transform can transform the response times of a fuel-cell O_2 analyzer to approximate that of the intrinsically faster responding paramagnetic or even zirconia-cell O_2 analyzers. This is helped by the low noise content of high-quality fuel-cell O_2 analyzers, which minimizes artifacts in the derivatized component. Figure 8.4 shows an example in which a zirconia-cell analyzer is compared to a fuel-cell analyzer in responding to injections of nitrogen into a flow-through system.

Some investigators (e.g., Frappell et al., 1989) have agonized over the disparity, which may be quite significant, between the effective volume of the system, which can be determined graphically from the slope of log-transformed oxygen concentration against time and the real chamber volume of the system. This loses sight of the fact that the effective volume of the system is a portmanteau term that includes not only the physical chamber volume of the system but also its flow characteristics, the length, diameter, and type of tubing, the volume and configuration of scrubbers, the response kinetics of the gas analyzers, and so on. It is best determined by iterative, empirical calculation, not from the actual volume of the system. There are more important things to worry about, such as the fact that the mass of the universe is proportional to its radius, not its volume.

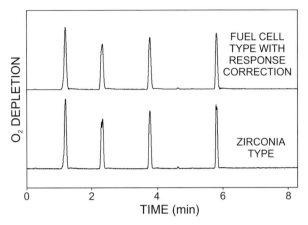

Figure 8.4 The instantaneous transformation can essentially eliminate any difference in response speed between fuel-cell O_2 analyzers and zirconia-cell O_2 analyzers, even with rapidly changing data. The data shown are from a good-quality fuel-cell analyzer (Sable Systems FC-10) and a zirconia-cell analyzer (original AEI S-3A). The peaks in O_2 depletion (presented here as positive-going; i.e., $F_iO_2 - F_eO_2$) were caused by 15-mL injections of nitrogen into a 1 L min^{-1} air stream flowing through both analyzers. The recording was lag-corrected, and data from the fuel-cell analyzer were subjected to instantaneous response correction in ExpeData. Peak volumes measured by integration agree within 0.1 percent between the two analyzers.

KNOW YOUR ANALYZERS

Gas analyzers are the eyes and ears of your respirometric research, and you should understand how they operate (Chapter 16). Know their strengths and limitations. For example, do not allow liquid water or volatile organic compounds to enter a zirconia-cell O_2 analyzer; do not expect a fuel-cell O_2 analyzer to deliver good results if its replaceable O_2 sensor is more than ~ 2 years old; and do not expect stable results from a paramagnetic O_2 analyzer if its sampling flow rate is unstable.

OXYGEN VERSUS CO$_2$ ANALYZERS

Because O_2-based flow-through respirometry must detect small changes in a very large signal (the O_2 content of the air, nominally at $\pm 209{,}939.2$ ppm if dried and STP-corrected; Tohjima et al., 2005), it demands a great deal from its analyzers. As covered in previous chapters, the most important factors in your success are:

- A high-quality O_2 analyzer; for the most demanding work a differential or dual-absolute O_2 analyzer is recommended.
- Setting the analog output of the analyzer, if applicable, to the smallest range that will cover the anticipated signal range. Zero-suppressed ranges such as 19–21.5 percent O_2 or (better still) digital data transfer are strongly recommended.

- Ensuring that the incurrent O_2 concentration is constant by using outside air or buffering the incurrent air with a carboy (see Figure 4.2). For critically sensitive work, scrubbing the incurrent air of water vapor and CO_2 is recommended.
- Averaging the analog signal from the analyzer, if applicable, during data acquisition.
- Maintaining the O_2 analyzer in the most stable environment possible. Leaving it turned on during measurement-intensive seasons is recommended.
- Finally, calibration. This subject is covered in detail in Chapter 16 and is important reading for anyone using O_2 analyzers.

Baselining can range from occasional to obsessive-compulsive, depending on the size of the $F_iO_2 - F_eO_2$ signal, but it must be done. A poor baseline will render your measurements unpublishable. *Always design your respirometry protocols with the central importance of baselining in mind.*

CO_2 analyzers are far more forgiving because they respond to upward changes from a small or zero baseline (i.e., F_iCO_2). In addition, because of this, F_eH_2O has a far less-significant effect. In the case of O_2, F_eH_2O dilutes the standing pool of oxygen in the air, pushing down F_eO_2 very significantly in most cases, often comparably to or even in excess of the metabolic signal. F_eH_2O dilutes CO_2 as well, but it causes only a small reduction in the magnitude of F_eCO_2 rather than a massive artificial signal.

In most cases, to obtain the most accurate results you will need to measure both O_2 and CO_2 (but see equation 9.16 and accompanying discussion). If you want to measure only one, then O_2 is better because (1) it is more widely used and reported, (2) it is more easily and accurately translated to energy metabolism units such as Watts or kilocalories per hour[1] (see especially equation 9.16), and (3) it is less prone to give false signals in the presence of an acid–base imbalance in the organism being measured. If your organism is very small, however (less than about 100 mg for an insect at room temperature, say), you will either need to abandon flow-through respirometry for closed-system or stop-flow respirometry (Chapter 4), use an expensive differential O_2 analyzer and use it well, or move to CO_2 respirometry, which is intrinsically far more sensitive. The chief disadvantage of using CO_2 as your sole flow-through respirometry gas metric is that its energy equivalence varies with RQ over a wide range; thus, to measure energy metabolism using only CO_2, you should ideally know the organism's RQ. You will also need to be able to convince referees that the organism was not exerting itself during the measurements to the point where lactate buildup occurs, which drives the CO_2–bicarbonate equilibrium toward CO_2, resulting in a large CO_2 signal that is chemical rather than metabolic in origin. This can be a significant drawback with poorly aerobic creatures such as reptiles and amphibians.

KNOW YOUR DATA ACQUISITION SYSTEM

Use the best and highest resolution system you can afford. By all means use programing languages to create your own system if you have the necessary free time; otherwise, you will be more productive if you use a flexible, commercially available system. Do not be blinded by fast sampling speeds; they can be useful, but for most research in respirometry, excessively fast data acquisition systems are a liability because they will pick up fast, transient noise signals that are ignored by slower systems. Make sure that the system supports intrasample averaging. Be careful, if you intend to use the system for automatic baselining or switching between respirometry chambers, to find out whether the data acquisition system contains built-in support for controlling instrumentation, preferably with flexible positioning of the control events during recording. Most packages intended primarily for teaching cannot do this easily or at all. Also check whether the manufacturer of the system in question supports its use in a research environment; some companies, such as Vernier Software, produce products that are excellent for basic education or hobbyist use, but they specifically prohibit the use of their products for research. Ask friends and colleagues about the usability of systems and, most important, about the expertise and the academic qualifications of the technical staff supporting the system's users.

A few general tips for getting the best results from your data acquisition system are in order.

- Use the highest resolution analog-to-digital converter available. Resolution is far more important than speed; 12 bits is usually adequate but may not be enough for some purposes.

- If available, use differential inputs. Consult the system's documentation on how to use them.

- Ensure that the system is capable of averaging voltage signals (taking a settable number of subsamples and storing only the mean or median, depending on the system).

- Set your analyzers to their smallest range that still covers the signals of interest. For example, set an O_2 analyzer to its 19–21.5 percent range, if available, or a CO_2 analyzer to a 0–1 percent range for a vertebrate.

- If you can sidestep analog data acquisition by using direct digital transfer of data, do so without hesitation.

- Record everything you reasonably can, including variables you assume will stay constant.

- Choose a system that allows you to save frequently used recording setups for quick recall.

- Do not calculate respirometry variables such as rate of O_2 consumption in real time unless you urgently require feedback during the recording. Calculate them during data analysis, after baseline and time-lag correction. If you nevertheless want to calculate them in real time, do so, but use the resulting data for quick checks only; do not use the data for publication.

- Comprehensive data analysis facilities are a must. Flexible data transformation ability is essential.

- The ability to record and play back data analysis macros will save you a great deal of time and is almost essential.

Generating a flow rate

For low flow rates (up to a few liters per minute), diaphragm pumps give good results and have the advantage that, ideally, the gas being pumped through them does not intermix with the surrounding air. Small diaphragm pumps can be powered from 12-V DC and can be used for either lab or field applications. Good-quality pumps that do not fail catastrophically early in their use are quite expensive. Ball-bearing rather than sleeve-bearing motors should be used. Flow rates can be set by altering pump drive (see below) and by using a special variable-orifice valve designed to give smooth flow control, referred to as a needle valve (see Appendix 1). The pump drive, and thus the flow rate, can be adjusted by changing the voltage applied to the pump, or, more ideally, by pulse-width-modulating the voltage applied to it. Pulse-width modulation (PWM) allows the pump to run at much higher torque at low speeds than merely running the pump at a low voltage. The PWM frequency should be ultrasonic or the motor will whine annoyingly. Integrated instruments containing a PWM-drive diaphragm pump, mass flow meter, and needle valve are available and are suitable for generating flow up to about 5 L min^{-1}.

For the best accuracy, a good mass flow meter is essential, and the best overall results are usually obtained with a mass flow control valve, which is a "set and forget" device that offers excellent accuracy and a constant flow rate (see Chapter 17).

For higher flow rates, a more brutish diaphragm pump can be used, but at flow rates greater than about 20 L min^{-1}, a centrifugal blower pump becomes the most practical flow driver. They are quiet and highly reliable. A variable-drive blower motor can be used in conjunction with a high-flow mass flow meter; some integrated systems are available that maintain a constant flow rate by closed-loop control. It is even possible, if events dictate it, to hook a shop vacuum up to a variable motor controller and create a cheap and cheerful, if noisy, high-flow pump.

At flow rates greater than about 200–1000 mL min^{-1}, depending on the analyzers, it is not practical or desirable to flow the sample through the analyzers (and scrubbers, if any). There are several reasons for this. Some O_2 analyzers, notably paramagnetics, can be mechanically disturbed and even damaged by high flow rates. Zirconia-cell analyzers can exhibit problems from cell cooling. All analyzers are sensitive to the pressure required to generate such rapid flow and will interpret it as an altered O_2 concentration. This is the case because the analyzers actually measure not O_2 concentration but O_2 partial pressure, as explained in Chapter 16. This effect becomes more and more significant as flow increases and is best avoided. As far as scrubbers are concerned, they wear out and will need to be replaced or replenished proportionately more rapidly as the flow rate through them increases (Chapter 19).

There is seldom any need for the flow rate through the analyzers to exceed about 200 mL min^{-1}. Thus, many respirometry systems subsample the main flow rate through the chamber, sipping a small aliquot for analysis. The subsampling flow rate is not critical, though for best results it should at least be stable, and it should be recorded together with other system variables. The only flow rate that needs to be known with good accuracy is the primary flow rate through the chamber or mask.

It is critical when subsampling to ensure that the gas sample is taken without transferring any significant pressure from the main gas flow. This is best done by flowing the main air stream into a manifold, such as an open-ended piece of tubing, a pipe, a T-adapter, or a syringe barrel (see Figure 16.1), from which the subsample can be pulled.

CRAZY PLUMBING

There are a few flow-plumbing bloopers that should be absolutely avoided. Correct plumbing for flow-through respirometry is covered in Chapters 10 and 11. Here, I present a few no-nos. Most important: never expose a gas analyzer to anything other than near-atmospheric pressure; never, for example, position a flow restrictor such as a needle valve upstream from an analyzer and a pump downstream from it (Figure 8.5): It will give completely incorrect readings. *Areas of unusually high or low pressure should only occur between a pump and a needle valve or mass flow control valve and should never include a respirometer chamber, a scrubber column, or a gas analyzer.* Other pitfalls include attempting to push a flow into a system while pulling it from the other end (Figure 8.5). Use common sense.

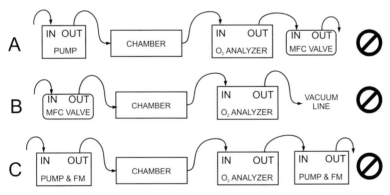

Figure 8.5 How not to plumb a respirometry system; three ways to guarantee at best failure and at worst catastrophe. (A) Push system using mass flow control valve (MFC); a needle valve plus rotameter could replace MFC. A high pressure is required to drive flow through either, causing elevated pressure in the chamber and analyzer. This distorts readings and may blow up the chamber or damage the analyzer. (B) Pull system using house vacuum (a pump can replace the vacuum line). Flow control is implemented ahead of the chamber and analyzer, resulting in subatmospheric pressure in both. Readings will be distorted, and the experimental animal may be harmed. (C) The experimenter has implemented a push system, then forgotten about it and plumbed a pull system as well (more than once I have seen this happen). Distorted readings are guaranteed. Depending on the relative strengths of the flow systems, a vacuum or high pressure may harm the experimental animal and/or blow up the chamber and analyzer.

One somewhat un-obvious point should be raised here. If you are pulling a subsampled flow from a main air stream, you will likely use a small diaphragm pump to do so. Moreover, you may use a needle valve to control the subsampled flow rate. If you do so, position the needle valve before, rather than after, the pump. If you position the needle valve after the pump, high-pressure pulsations will occur in the plumbing separating the output of the pump from the input of the needle valve. If the water vapor partial pressure in that plumbing line exceeds the value corresponding to dewpoint at the ambient temperature, liquid water will condense in it. It can then dribble into your desiccants (if any) and analyzers. You have been warned.

HIGH FLOW RATES: COMPENSATING FOR WATER VAPOR

Given that subsampling the excurrent air helps reduce the use of scrubber chemicals, some readers may object that if a high flow rate is measured and if it contains a significant amount of water vapor, and if that water vapor is subsequently scrubbed from the subsampled air stream before analysis, the primary flow rate is in fact overestimated by the amount of water vapor present in it. This is quite correct and is easy to compensate for provided that the partial pressure of water vapor in the air stream is known or can be calculated, and that the ambient barometric pressure is known. If water vapor pressure is WVP and barometric pressure in the same units is BP, the dry-corrected flow rate, FR_c, is

$$FR_c = FR\left(BP - WVP\right)/BP \tag{8.6}$$

where FR is the uncorrected flow rate. See the section "Mathematical scrubbing" for advice on measuring WVP and BP, and see Chapter 16 for methods of calibrating water vapor analyzers.

MEASURING FLOW RATES

Chapter 17 is devoted to measuring flow rates and includes much essential information, including techniques for calibrating flow systems. In all flow-through respirometry, the primary flow must be measured and recorded continuously during each experiment. If you are using a rotameter, invest instead in a mass flow meter or controller. In some very high flow rate systems, a suitable mass flow meter might not be easily feasible, and in that case, nitrogen dilution calibrations will have to be performed, preferably after each run, as described in Chapter 17.

If you are subsampling, you should if at all possible record your subsampling flow rate as well. Good-quality integrated subsampling units include flow meter analog outputs. If the system's response becomes sluggish, these data allow you to quickly determine, for example, whether a scrubber column is becoming clogged or a pump diaphragm is starting to wear out. It is far better to record too many data than to record too few. In fact, some investigators (including yours truly), when operating a sealed respirometry system, will neurotically measure flow rates at the inlet and outlet of the system through-out an experiment. Any leak is immediately obvious. Nowadays, RAM and storage can be regarded as infinite. Use them to the full.

SCRUBBING THE AIR STREAM

It is customary (though not essential, if all of the necessary parameters (O_2, CO_2, and water vapor concentration) are measured) to scrub either water vapor or water vapor and CO_2 from the sampled air stream, and in some cases from the incurrent air stream as well. The techniques for doing this, and the characteristics of the different techniques, together with information on tubing and connectors, are covered in Chapter 19.

Mathematical scrubbing

Of particular interest is the recent tendency to eliminate all scrubbers and compensate for the dilution effect of water vapor by using Dalton's law of partial pressures (e.g., Lighton, 2008; Melanson et al., 2010). If WVP and BP are known (in any units), then any gas species of concentration X present in the mixture of gases with total pressure BP can be converted to its "dry" equivalent X' thus:

$$X' = X \times BP/(BP - WVP) \qquad (8.7)$$

Using this approach allows you to dispense with the water vapor scrubber entirely. To do so, you need to know $F_e H_2 O$, which is most easily measured as the partial pressure of water vapor divided by barometric pressure (also known as the sum of all partial pressures). Thus, if you decide to measure water vapor and compensate for its dilution effect, you will need to measure *both* kilopascals of WVP and kilopascals of total BP (equation 8.7).

Good-quality microcontroller-based water vapor analyzers can display water vapor pressure directly. If one is not available, then measure the relative humidity of the air-stream and the temperature at which the relative humidity was measured. Calculate the kilopascals of saturated water vapor pressure at that temperature (equation 16.7) and multiply that figure by measured relative humidity divided by 100 to get water vapor pres-sure (equation 16.6). Some good-quality gas analyzers have BP outputs, making record-ings of this variable easy. You will need to record all the relevant variables that will give you access to WVP and BP during the recording. Then, before performing any analysis on the O_2 channel, it should be multiplied by $BP/(BP - WVP)$, which is equivalent to dividing by $(1 - F_e H_2 O)$; see equation 8.7. This will compensate for the water vapor dilution effect if the water vapor analyzer is properly calibrated, and if the signals from the water vapor and O_2 analyzers are lag-corrected so that they are in phase. Especially in the case of O_2, bear in mind that the analyzer should, where possible, be calibrated against dry air, so that

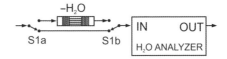

Figure 8.6 Switching a desiccant column in or out of an excurrent air stream for water vapor dilution validation. S1a and S1b are electrically operated gas switching solenoid valves and are switched simultaneously under manual or automatic control. If water vapor correction is implemented correctly, switching the desiccant into the flow stream should have no steady-state effect on dilution-compensated data. If switching in the desiccant does affect the data, then the disparity can be used to calibrate (span) the water vapor analyzer so that the disparity is eliminated.

water vapor dilution compensation does not affect the accuracy of the analyzer. If the analyzer was calibrated to 20.94 percent O_2 in moist air, then compensating for the dilution caused by that water vapor later will lead to an overestimate of O_2. This can be avoided by inserting a small desiccant column before the O_2 analyzer when calibrating it. In fully automatic systems this can be done programatically, using a pair of changeover solenoid valves to select or bypass the calibration-only desiccant column (Figure 8.6).

Given that an O_2 analyzer will measure F_iO_2 when the solenoid pair is idle and F'_iO_2 when it is energized, such a system also provides an effective way to calibrate a water vapor analyzer at its zero and span points. This is best done with a stable source of incurrent air, a carefully calibrated O_2 analyzer, a top-grade desiccant such as magnesium perchlorate or Drierite, and a water vapor analyzer with a direct readout in partial pressure of water vapor. First, the solenoid pair are energized, F'_iO_2 is measured, and the O_2 analyzer is spanned appropriately (see Chapter 16 and especially Tohjima et al., 2005). At the same time, the water vapor pressure analyzer is zeroed. Then the solenoid pair are de-energized and F_iO_2 is measured. It can be shown that

$$\mathrm{WVP} = \mathrm{BP}\left(F'_iO_2 - F_iO_2\right)/F'_iO_2 \qquad (8.8)$$

where WVP is in the same units as BP, and the other terms are as previously defined. The water vapor analyzer is then spanned to the calculated value of WVP. If the water vapor analyzer reads in RH%, it will need to be spanned to the calculated relative humidity at the temperature of the analyzer (see equations 16.6 and 16.7). Water vapor pressure analyzers are covered in Chapter 16.

Mathematical compensation is by far the most elegant method for water vapor dilution compensation and works very well in practice. So well, in fact, that in my opinion it significantly outperforms other means of "compensating" for water vapor dilution by attempting to remove it. For example, thermal water vapor scrubbers are ineffective at temperatures below 1–2°C at best, limiting the WVP they can remove from the airstream to approximately 0.7 kPa, corresponding to a dewpoint of that temperature; some even work at 5°C, producing "dried" air with a whopping 0.9 kPa WVP. This can cause significant errors (Melanson et al., 2010). An example of actual WVPs that will cause failure of even maximally efficient thermal scrubbing but will have no effect on mathematical water vapor dilution compensation is shown in Figure 8.7.

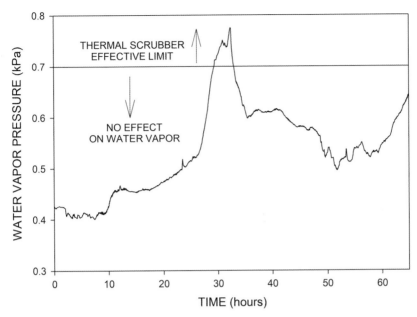

Figure 8.7 Water vapor pressure (incurrent values) measured in a metabolic phenotyping system. The horizontal line denotes the water vapor removal limit of a maximally efficient thermal scrubber, corresponding to approximately 0.7 kPa (most such scrubbers only "dry" to 0.9 kPa and would have had no effect at all). The actual water vapor pressure remains well below that limit for most of the recording; the thermal scrubber will thus have had no effect during those periods. This will have caused serious inaccuracy if a thermal scrubber had been used in this system. Fortunately, this system utilized direct mathematical compensation for water vapor dilution, using the measured values of water vapor pressure rather than assuming (incorrectly) that water vapor was removed. (Data used by kind permission of David Wasserman and Owen McGuinness, Vanderbilt Mouse Metabolic Phenotyping Center (MMPC).)

Because I believe that the mathematical compensation for water vapor dilution is an important technique with wide-ranging benefits in the field of metabolic measurement, I will now guide you through a validation. This is necessary because there exists much fear, uncertainty, and doubt regarding the validity of this technique, most of it generated by people unable or unwilling to use it. To this end, I am using raw data from the Vanderbilt Mouse Metabolic Phenotyping Center with the kind permission of David Wasserman and Owen McGuinness.

First, I extract all of the incurrent O_2 values for one O_2 analyzer from a recording lasting 74 h (Figure 8.8, solid line). The O_2 analyzer was spanned to 20.95 percent STPD before the recording began. As you can see, the raw O_2 values are quite low and variable, reflecting the fact that BP is variable too, and below the standard value of 101.325 kPa (Figure 8.8, dashed line). In fact, the extent of incurrent O_2 variability is comparable to the O_2 depletion caused by a mouse at the system flow rate (2 L min^{-1})! Incurrent O_2 should be constant (Tohjima et al., 2005); from what does this variability arise? Perhaps

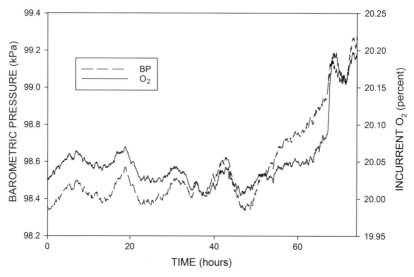

Figure 8.8 Approximately 74 h of incurrent O_2 data (solid line) and BP data (dashed line) from a metabolic phenotyping system. The O_2 analyzer was calibrated to 20.95 percent before the start of the recording, and its low and highly variable readings reflect the influence of BP and water vapor dilution (see text).

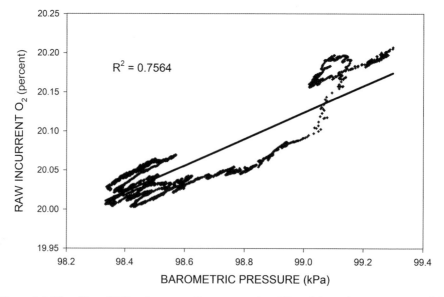

Figure 8.9 The effect of BP on incurrent O_2 concentration. BP explains only ~ 75 percent of O_2 variance, so something besides BP is affecting incurrent O_2 concentrations (see text).

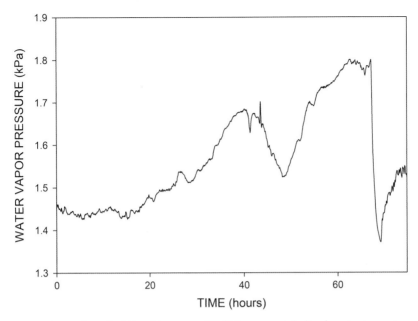

Figure 8.10 Approximately 74 h of incurrent WVP from a metabolic phenotyping system (see Figure 8.8 for O_2 and BP data from the same system).

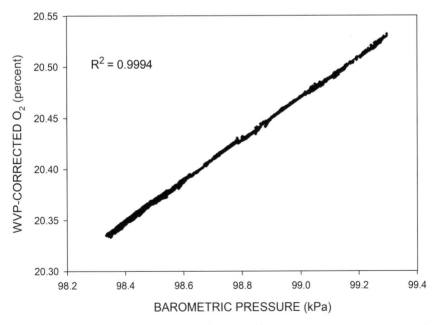

Figure 8.11 After correcting the raw O_2 data for WVP dilution using equation 8.7, BP variations now explain 99.94 percent of O_2 variance.

BP? The O_2 and BP traces are obviously highly correlated, though this correlation breaks down somewhat at around 60 h.

We can determine how much of the variability in O_2 concentration was caused by variations in BP using linear regression analysis (Figure 8.9)—only ~ 75 percent, leaving ~ 25 percent unexplained.

Plainly, something other than BP is affecting the O_2 signal. Could it be WVP? Let us examine the incurrent WVP over the same timespan (Figure 8.10).

Let us now apply equation 8.7 to the raw O_2 data, thus removing the dilution effect of WVP on O_2 by using the WVP and BP data in accordance with Dalton's law of partial pressures, and then test the effect of BP variation again (Figure 8.11).

As you can see, our unexplained variance has decreased from ~ 25 percent to ~ 0.06 percent after compensating for the dilution effect of water vapor. Now, correcting the raw O_2 trace for both BP and water vapor dilution, we find that the incurrent O_2 concentration is far more stable (Figure 8.12, upper trace). In fact, it hardly deviates at all from the analyzer's original calibrated value of 20.95 percent. Nevertheless, we know that precise baseline measurements were made every 2.5 min in this system, and we can use those to calibrate the O_2 analyzer to 20.94 percent with each baseline, reducing analyzer error to almost zero. This assumes that incurrent O_2 concentrations after correction for BP and WVP are constant, which is justified (Tohjima et al., 2005).

The result is practically perfect incurrent O_2 concentration measurements with an SD of only 0.0003 percent or 3 ppm (Figure 8.12, lower trace), thus ensuring that O_2 depletions

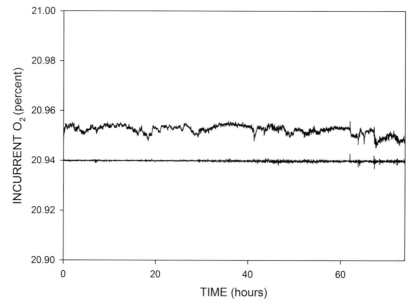

Figure 8.12 Incurrent O_2 concentrations after correction for BP and WVP (top trace). The O_2 analyzer was originally calibrated to 20.95 percent. The SD over the entire 74-h period without further correction is only 0.002 percent. After correction to an incurrent O_2 concentration of 20.94 percent at each baseline (bottom trace), the SD falls to 0.0003 percent.

caused by the experimental animals will be extremely precisely measured—without the myriad problems caused by sample air desiccators of any kind.

EQUATIONS

The exact equation to use for transforming your gas analyzer readings obtained via flow-through respirometry into meaningful results is a source of great confusion. See Chapter 9 for the background and Chapters 10 and 11 for specifics regarding systems where the air is pushed or pulled, respectively, through the chamber containing the animal.

THE RESPIROMETRY CHECKLIST

Here are a few suggested items to consider carefully before starting a flow-through res-pirometry experiment, which may be part of a larger project or may constitute the whole project. Even veterans will save time and frustration in the long term by setting aside a few minutes at the start of a new project for a systematic evaluation. One prereq-uisite is simple: tools. If you don't already have the following, get them, because at some point you'll need them. They are available from electronic distributors (see Appendix 1). Do not skimp on cost: Get the highest quality your lab can afford, rather than dime-store knock-offs made from salvage-metal alloys with the structural strength of parmesan.

- A set of screwdrivers, including slot, Philips, and Torx heads, in a reasonable range of sizes.
- A set of pliers and diagonal cutters, small and large, plus at least one vice-grip plier.
- Scissors, box-cutter, Exacto knife set, and single-edge razor blades.
- A set of hex keys and wrenches, metric and standard.
- A good-quality multimeter. Good brands include Fluke and Agilent. Acceptable economy brands include B&K Precision and Wavetek. Cheap multimeters (like cheap instruments in general) are unreliable and inaccurate.

With these tools at hand, you'll save a lot of time and frustration. Don't think you can do without them; you can't. Keep them in a secure place, or you will replace them repeatedly until they saturate their binding sites in other labs.

The checklist, in approximate order, is as follows:

Be clear about your goals. Write down the goals of the experiment. Do the goals dictate the form of the system? For example, must RQ be determined, in which case is both O_2 and CO_2 analysis required? Is direct measurement of RQ practical in flow-through mode, or should it (for very small animals such as *Drosophila*) be done separately in groups of animals or on individuals using constant volume techniques (Chapter 4)? What speed of response is required? For example, treadmill energetics of small organisms may involve tradeoffs between chamber volume, analyzer noise and drift, and flow rate.

Push or pull? Decide which flow-through mode you will use (see Figure 4.7). Often your goals will dictate which mode you will use. For example, a practical mask system operates only in pull mode.

Look at the equations. Respirometry equations differ considerably between push and pull systems (Chapter 9). For example, scrubbing CO_2 before O_2 analysis can greatly simplify your calculations. Decide in advance which equation you'll use.

Decide on scrubbers. Chemical scrubbers are often required before gas analysis (especially O_2 analysis). Do you need to use them? If so, what kind (Chapter 19)? Can you do without scrubbers for at least part of the flow path, especially a high-flow pull path that might have an insatiable appetite for expensive scrubbers, by compensating mathematically for water vapor dilution (equations 8.6 and 8.7)?

Draw out the system. Drawing a detailed diagram of the system will force you to think through the system's design. Plus, it is a useful item of documentation and will save you and your students or colleagues time when similar setups are needed in the future.

Choose your tubing and connectors (see Chapter 19). From your diagram, you'll be able to estimate what you need.

List data acquisition channels. Make a list of all of the variables you'll record, such as O_2 concentration, barometric pressure, and ambient temperature. Assuming that you're using a voltage-based data acquisition system (almost universal, though becoming less so), write down the analog output range of each instrument and the transfer equation that relates its voltage output to the units it is measuring. Also decide which channels you'll filter for the best noise reduction. This list is an important planning tool and independent record, even if your data acquisition system supports saving setup files that can instantly reconfigure it to a previous setup. It is also an important part of leaving an audit trail, allowing you to return to your collected data and understand them months or years later. Add the electrical connections to your system diagram, or make a separate diagram for them.

List instrument control events. If you're using automated baseline or switching between channels, now is the time to plan your strategy, which will vary between data acquisition systems.

Check everything during assembly. To the extent possible, check whether every instrument is functioning as intended before deploying the system as a whole. Make electrical connections and check that the data acquisition system is getting sensible readings. Slowly assemble the system, double-checking everything against your system diagrams as you go. Think in modular terms as much as possible. Don't proceed to the next step until the step you're working on is complete and functional. *The classic mark of a beginner is to assemble a complex system without checking anything and then, when it doesn't work, panic and assume that the instruments are faulty.*

Use the scientific method to solve problems. If something isn't working as expected, isolate the problem by identifying and then testing each probable cause of the problem. For example, if an analyzer's reading doesn't change, is the flow system (both main flow and subsample flow, if applicable) working? Is a scrubber column blocked? Is there a leak in the system? As a rule, four-fifths of problems are caused by incorrect or faulty connections, whether plumbing or electrical; nearly one-fifth are caused by incorrect assumptions or calibrations; and 1 percent or fewer are caused by equipment issues.

Don't be shy about asking for help. Ask your more experienced colleagues for help even if you found their behavior at the last faculty meeting to be annoying. Contact the vendor of your instruments or system, assuming (which is not always the case, by any means) that they know what they are doing. Whatever course of action you take, however, don't simply announce that the system doesn't work and leave it at that. Accurately describe the problem and, if possible, the steps you've already taken to diagnose and fix it. If feasible, include plumbing diagrams. Simply saying "It doesn't work" will not help anyone, least of all you.

9

Flow-through Respirometry: The Equations

THE EQUATIONS

No aspect of flow-through respirometry induces more fear and loathing than its equations. This is partly because most published accounts compliment the reader's intelligence by leaving out numerous steps in the derivations, or assuming that the reader has, like the paper's author, spent years thinking these things through. My goal in this chapter is to write an account that includes enough detail to allow you to choose the equation you need for any likely setup you may put together, while hopefully making it clear why that equation works under that circumstance.

The key to understanding the equations is to think for a moment, paradoxically, about nitrogen. Nitrogen is the nonparticipant in respirometry. If we flow air past a specimen in a chamber, that specimen will not affect the amount of nitrogen in the air stream, which will be the same in the air entering and the air leaving the chamber. The gases the specimen *will* affect are O_2, CO_2, and water vapor.

Now let us assume that we are measuring the flow rate of air entering the chamber. We call that flow rate FR_i and it consists of the incurrent flow rates of all of the separate gas species (nitrogen, oxygen, CO_2, and water vapor). We denote the fractional concentrations of the incurrent non-nitrogen gas species F_iO_2, F_iCO_2, and F_iH_2O. F_iO_2 will usually be somewhere around 0.208 (0.2094 for outdoor air if F_iCO_2 and F_iH_2O are zero). F_iCO_2 will usually be about 0.0005, and F_iH_2O will be about 0.01 in a typical building. We know that the flow rates of nitrogen into and out of the chamber are exactly equivalent. The flow rate of nitrogen into the chamber is $FR_i(1 - F_iO_2 - F_iCO_2 - F_iH_2O)$. Likewise, we denote the fractional concentrations of the excurrent non-nitrogen gas species as F_eO_2, F_eCO_2, and F_eH_2O, and the overall flow rate leaving the chamber as FR_e. The rate at which nitrogen flows from the chamber is $FR_e(1 - F_eO_2 - F_eCO_2 - F_eH_2O)$. The keystone to this logic is the fact that these two expressions are exactly equivalent, so

$$FR_i\left(1 - F_iO_2 - F_iCO_2 - F_iH_2O\right) = FR_e\left(1 - F_eO_2 - F_eCO_2 - F_eH_2O\right) \qquad (9.1)$$

Thus, it follows that

Measuring Metabolic Rates: A Manual for Scientists. Second Edition. John R. B. Lighton, Oxford University Press (2019).
© John R. B. Lighton (2019). DOI: 10.1093/oso/9780198830399.001.0001

$$\text{FR}_i = \text{FR}_e \left(1 - F_e O_2 - F_e CO_2 - F_e H_2 O\right) / \left(1 - F_i O_2 - F_i CO_2 - F_i H_2 O\right) \qquad (9.2)$$

And, of course,

$$\text{FR}_e = \text{FR}_i \left(1 - F_i O_2 - F_i CO_2 - F_i H_2 O\right) / \left(1 - F_e O_2 - F_e CO_2 - F_e H_2 O\right) \qquad (9.3)$$

You can, in principle, measure all of these concentrations, both incurrent and excurrent, if you have the appropriate analyzers. Optionally, incurrent water vapor and CO_2 can be scrubbed down to zero. The incurrent concentrations (whether zero or not) are referred to as the *baseline* values and are usually measured with no animal in the chamber or, if you are clever and suitably equipped, by using a manually operated or computer-controlled air valve to direct either incurrent or excurrent gas concentrations through the analyzers. For any conventional organism not engaging in photosynthesis, $F_e O_2$ will be lower and $F_e CO_2$ and $F_e H_2 O$ will be higher than their baseline or incurrent values.

How do we calculate $\dot{V}O_2$ (the rate of O_2 consumption), $\dot{V}CO_2$ (the rate of CO_2 production), and $\dot{V}H_2 O$ (rate of water vapor production) from these data? By definition, the rate of O_2 consumption is equal to the flow rate of O_2 in the incurrent air minus the excurrent O_2 flow rate. The reverse holds true for CO_2 and water vapor, which are produced rather than consumed. So,

$$\dot{V}O_2 = \text{FR}_i \left(F_i O_2\right) - \text{FR}_e \left(F_e O_2\right) \qquad (9.4)$$

And, analogously,

$$\dot{V}CO_2 = \text{FR}_e \left(F_e CO_2\right) - \text{FR}_i \left(F_i CO_2\right) \qquad (9.5)$$

and

$$\dot{V}H_2 O = \text{FR}_e \left(F_e H_2 O\right) - \text{FR}_i \left(F_i H_2 O\right) \qquad (9.6)$$

Now, to go further, it is of utmost importance that you decide whether you are measuring incurrent or excurrent mass flow (FR_i or FR_e); or usually, whether you are pushing or pulling air past the animal. Figure 4.7 illustrates the difference.

Let's examine the case where we know FR_i. In this case, we substitute $\text{FR}_i(1 - F_i O_2 - F_i CO_2 - F_i H_2 O)/(1 - F_e O_2 - F_e CO_2 - F_e H_2 O)$ for FR_e in equation 9.6. This is the rather messy result:

$$\dot{V}O_2 = \text{FR}_i \left\{ F_i O_2 - \left[F_e O_2 \left(1 - F_i O_2 - F_i CO_2 - F_i H_2 O\right) / \left(1 - F_e O_2 - F_e CO_2 - F_e H_2 O\right)\right]\right\} \qquad (9.7)$$

And, likewise,

$$\dot{V}CO_2 = \text{FR}_i \left\{ \left[F_e CO_2 \left(1 - F_i O_2 - F_i CO_2 - F_i H_2 O\right) / \left(1 - F_e O_2 - F_e CO_2 - F_e H_2 O\right)\right] - F_i CO_2 \right\} \qquad (9.8)$$

and

$$VH_2O = FR_i \left\{ \left[F_eH_2O \left(1 - F_iO_2 - F_iCO_2 - F_iH_2O \right) / \left(1 - F_eO_2 - F_eCO_2 - F_eH_2O \right) \right] - F_iH_2O \right\}$$

(9.9)

In all cases, because fractional concentrations are dimensionless, the rates of gas production or consumption take the units in which the flow rate is expressed—typically, depending on the body mass of the organism, milliliters per minute or liters per minute. A similar set of equations can easily be derived for the case in which the excurrent flow rate is known but the incurrent flow rate is not, as in a mask configuration. I return to this in a moment.

These equations appear quite complicated, but their concepts and derivations are very simple. In the coming chapters I often present simpler equations derived from the above, so that there is seldom a need to use these equations in their full form. Nevertheless, they are here if you want to use them.

In the case of water vapor, we are usually not interested in volumes of water vapor but in the mass of liquid water lost by the animal. One easy way to convert from a volume of water vapor to a mass of water is to take STP-corrected water vapor production rate (i.e., measured with a mass flow controller or, less ideally, corrected to STP from a rotameter reading) and multiply it by $18/22,414$ or 0.803 mg of water per milliliter of water vapor at STP. Thus, for example, a loss of 2.1 mL of H_2O vapor min^{-1} corresponds to losing a mass of 1.69 mg H_2O min^{-1} or 0.101 g H_2O h^{-1}.

From all of the above, it's evident that you really do need to be able to either put hard values on CO_2 and water vapor concentrations, or eliminate them from the chosen equation by scrubbing them—a strategy that is usually only partially successful, depending on where the particular gas species was scrubbed relative to where the flow was measured. As Withers (2001) points out in his excellent article on this subject, assuming that incurrent and excurrent flow rates are identical leads to the simple equation $VO_2 = FR_i(F_iO_2 - F_eO_2)$, an equation often seen, and often used without questioning, in the calculations of respirometry novices. When both CO_2 and water vapor are scrubbed from the analyzed air, this equation yields data about 25 percent too low in typical situations. This is a *very* common error—one that, in my capacity as a referee of scientific papers, I have seen many, many times.

What about measuring only FR_e, not FR_i as in equations 9.7–9.9? Here the situation becomes a bit more complicated because FR_e can be measured in three distinct ways:

- Straight up, with no scrubbers (FR_e)
- Scrubbed of water vapor (FR'_e)
- Scrubbed of water vapor and CO_2 (FR''_e).

FR_e will always be larger than FR'_e, which will always be larger than FR''_e. Figure 9.1 illustrates this concept. In practice, unscrubbed excurrent flow rate (FR_e) is usually the measured parameter for larger animals, especially in the mask configuration, because the quantity of scrubbing chemicals required to produce FR'_e and especially FR''_e may be

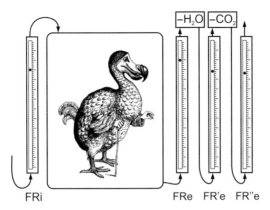

Figure 9.1 The relation between incurrent air flow rate (FR_i) and excurrent air flow rates with no scrubbing (FR_e), with H_2O scrubbing (FR'_e), and with both H_2O and CO_2 scrubbing (FR''_e).

prohibitively expensive and their presence may impede the high flow rates that are usually required.

Following the way Withers (2001) organized this set of choices and the resulting equations, we can create a hard-working group of eight equations that, together with the information presented above (especially equations 9.2 and 9.3 for interconverting FR_i and FR_e), will allow you to create a respirometry equation for practically any configuration of flow measurement and scrubber position. First we deal with the measurement of VO_2, VCO_2, and VH_2O:

$$VO_2 = FR_i\left(F_iO_2\right) - FR''_e(F''_eO_2) = FR_i\left(F_iO_2\right) - FR'_e(F'_eO_2) = FR_i\left(F_iO_2\right) - FR_e\left(F_eO_2\right)$$

$$VCO_2 = FR'_e\left(F'_eCO_2\right) - FR_i\left(F_iCO_2\right) = FR_e\left(F_eCO_2\right) - FR_i\left(F_iCO_2\right)$$

$$VH_2O = FR_e\left(F_eH_2O\right) - FR_i\left(F_iH_2O\right) \qquad .$$

(See also equations 9.7–9.9.) Next, we turn to methods for interconverting FR_e, FR'_e, and FR''_e:

$$FR'_e = FR_e\left(1 - F_eH_2O\right)$$

$$FR''_e = FR'_e(1 - F'_eCO_2) = FR_e\left(1 - F_eH_2O\right)\left(1 - F_eCO_2\right) \qquad .$$

Finally, use these equations to interconvert between the fractional gas concentrations:

$$F'_eCO_2 = F_eCO_2 / \left(1 - F_eH_2O\right)$$

$$F'_eO_2 = F_eO_2 / \left(1 - F_eH_2O\right)$$

$$F''_eO_2 = F_eO_2 / \left[\left(1 - F_eH_2O\right)\left(1 - F_eCO_2\right)\right] = F'_eO_2 / \left(1 - F'_eCO_2\right) \qquad .$$

You can now derive equations that allow the accurate measurement of VO_2, VCO_2, and VH_2O with any configuration of flow meter and scrubber. In the subsequent chapters I present "canned" equations, but above are the raw materials from which you can manufacture any respirometry equations you need.

Here is an example of a typical derivation. You wish to measure VO_2. You are measuring incurrent flow rate and are scrubbing CO_2 and water vapor from the excurrent *and* incurrent air. From equation 9.1, $VO_2 = FR_i(F_iO_2) - FR''_e(F''_eO_2)$. Substituting $FR_i(1 - F_iO_2 - F_iCO_2 - F_iH_2O)/(1 - F_eO_2 - F_eCO_2 - F_eH_2O)$ for FR_e (equation 9.3) and setting F_iCO_2, F_iH_2O, F_eCO_2, and F_eH_2O to zero because they are all scrubbed:

$$VO_2 = FR_i\left(F_iO_2\right) - FR_i\left[\left(1 - F_iO_2\right)/\left(1 - F''_eO_2\right)\right]\left(F''_eO_2\right) \tag{9.10}$$

Factoring:

$$VO_2 = FR_i\left\{F_iO_2 - F''_eO_2\left[\left(1 - F_iO_2\right)/\left(1 - F''_eO_2\right)\right]\right\} \tag{9.11}$$

this simplifies still further to the familiar

$$VO_2 = FR_i\left(F_iO_2 - F''_eO_2\right)/\left(1 - F''_eO_2\right) \tag{9.12}$$

which owes much of its popularity to the fact that VO_2 thus measured is independent of VCO_2 and thus of RQ.

Always be aware of the assumptions behind an equation and the practical effects of such assumptions. For example, equation 9.12 is dependent on the assumption that the air entering the chamber is dry and CO_2-free. If the incurrent air contains water vapor at a typical inner-building fractional concentration of about 0.01 (assuming you're at sea level, with a room temperature of $25\,°C$ and a relative humidity of 50 percent), then FR_i will be overestimated, relative to its dry state, by about 1 percent. The easiest way to compensate for this is to measure the water vapor pressure (directly or indirectly) and the ambient barometric pressure and correct the flow rate as shown in equation 8.6. FR_i can be corrected for F_iCO_2 by multiplying it by $(1 - F_iCO_2)$, but if the incurrent air contains CO_2 at a typical inner-building fractional concentration of < 0.001, then the effect of this correction may be negligible.

TRANSLATING TO ENERGY EQUIVALENTS

For many applications, it is necessary to transform gas exchange measurements into energy equivalents, most usually Watts or calories per hour. This can be done in various ways, but typically, the organism's RQ is plugged into an equation such as

$$\text{J mL O}_2^{-1} = 16 + 5.164\left(RQ\right) \tag{9.13}$$

(Lighton et al., 1987), thus yielding the oxyjoule equivalent per milliliter of O_2, and VO_2 is then multiplied by this conversion factor, yielding joules in the same time units as VO_2.

This is then divided by seconds per time unit to yield joules per second or Watts. For example, an animal with RQ = 0.83 has an oxyjoule equivalent of 20.29 J mL O_2^{-1}. If its VO_2 = 1.28 mL min^{-1}, then its energy expenditure is 25.97 J min^{-1} or 0.433 W. If other energy units are desired, they can easily be interconverted; for example, calories = joules × 0.239, so kilocalories (or calories) per day = Watts × 20.65.

The usual approach to converting gas exchange data to energy equivalents in the biomedical field is to use the Weir equation:

$$EE = 5.46 \times VO_2 + 1.75 \times VCO_2 \tag{9.14}$$

where EE is in kilocalories per day and VO_2 and VCO_2 are in milliliters per minute (Weir, 1949). Or, if you prefer,

$$EE = 0.06 \times \left(3.941 \times VO_2 + 1.106 \times VCO_2\right) \tag{9.15}$$

where EE is in kilocalories per hour, and VO_2 and VCO_2 are in milliliters per minute. Various forms of the Weir equation exist, in many different combinations of units, including forms that account for excreted nitrogen and are not, at present, within the scope of this book. Alternative, outdated equations for converting VO_2 and VCO_2 to EE, such as the Lusk equation (Lusk, 1928) are significantly less accurate than the Weir equation and should be avoided.

But here's an interesting fact: To convert gas exchange units to energy equivalents you do not necessarily need to know both VO_2 and VCO_2, or even guess at RQ. There's an interesting wrinkle first pointed out by Weir (1949), then rediscovered by Koteja (1996). If you are primarily interested not in VO_2 as such but in energy equivalents, then you can take advantage of the fact that the dependence of the oxyjoule equivalent on RQ is in the opposite direction to the error created when assuming a consensus RQ and using equations (such as equations 10.2 or 11.2) that are sensitive to RQ. Koteja (1996) showed that the error introduced by such equations does not exceed 0.6 percent over the entire physiologically meaningful RQ range, provided that an RQ of 0.8 is assumed and that the VO_2 figures are converted to energy equivalents. As Koteja said, "less accurate open-flow respirometric designs give better results" (p. 675) in this case, where energy metabolism is the primary measurement of interest. This is especially the case if VCO_2 is measured, because CO_2 calibration gases with an accuracy better than 1 percent are generally not available on the open market. Thus, the argument can be made that using this shortcut may often result in a more accurate measure of EE. Similar insights were reached by Arch et al. (2006).

Weir (1949) condensed his seminal insight into a succinct equation:

$$EE = 5\left(\Delta O_2\right)FR \tag{9.16}$$

where EE is in kilocalories per minute, FR is excurrent or incurrent flow rate through the chamber in liters per minute, and ΔO_2 is the fractional depletion of O_2 in the excurrent air stream. A rigorous derivation of this equation from first principles has confirmed its validity at normoxia, and verified its accuracy using empirical data (Kaiyala et al., in press, PLoS One). It can be easily modified to accommodate different EE and flow rate units.

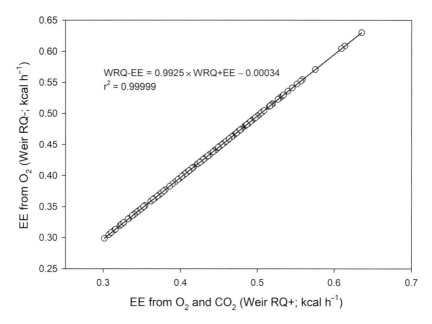

Figure 9.2 EE calculated via VO_2 and VCO_2 using the Weir RQ+ equation (X axis) and calculated from O_2 depletion alone using the Weir RQ– equation 9.16 (Y axis) are compared. Data are from eight C57BL/6J mice, 24 h per data point per mouse, at six temperatures ranging from 19 to 29°C over a 21-day period, using a Promethion metabolic phenotyping system. There were no outliers. If the two equations gave identical results, the slope would be 1.00. (Data are from Kaiyala et al., in press, PLoS One.)

How well does this equation (which Karl Kaiyala and I have nicknamed the Weir RQ- because it is independent of RQ) compare with data derived from both VO_2 and VCO_2 using the Weir equation? Exceedingly well, as it turns out (Figure 9.2).

The ~0.75 percent disparity between the two transformations $(100(1 - 0.9925))$ is within the 1 percent calibration uncertainty of the CO_2 span gas and close to the theoretically expected ~ 0.6 percent error of the Weir RQ– equation over the physiological RQ range. O_2 calibration is unlikely to introduce significant errors (see Chapter 16). This leads me to suggest that if EE alone is of primary importance, measuring VCO_2 is optional or, perhaps, unnecessary. It could even lead to less accurate EEs, considering the limited precision of commercially available CO_2 calibration gases—not to mention the proclivity of calibration gas manufacturers to produce poorly verified mixtures.

If you are concerned about losing RQ data, consider that by Hess's law of constant heat sums, mean food quotients and RQs are equivalent over entire circadian cycles in an animal at body mass equilibrium (see Chapters 13 and 14). Thus, even measuring VCO_2 and thus RQ may not be necessary unless short-term fluctuations in metabolic substrate utilization are of interest or the animal is subjected to physiological states such as prolonged overeating (de novo lipogenesis resulting in RQ or RER > 1) or fasting (ketosis resulting in RQ or RER < 0.7).

10

Flow-through Respirometry: Incurrent Flow Measurement

In systems measuring incurrent flow, air is usually pushed (though in principle it can be pulled) through a chamber, and the metabolic rate of an organism within the chamber is derived from the concentration changes it causes between incurrent and excurrent concentrations of O_2 and CO_2. Flow rates may vary depending on the size and type of animal being measured, between a few milliliters per minute and hundreds of liters per minute. As a matter of practicality, most respirometry systems for large animals are of the pull, rather than push, type, with downstream flow metering (see Chapter 11).

The mechanics of pushing the air through the chamber at a known flow rate varies with the size and type of the chamber and with any requirements for pretreating the air. For small animals such as insects, it is not only practical but in most cases necessary to scrub CO_2 from the incurrent air to yield a stable baseline for CO_2 emission measurement. The same goes for water vapor measurement, although for many insects, low water vapor pressure will cause physiological stress and even behavioral reactions (Lighton and Bartholomew, 1988; Gibbs et al., 2003). In such cases, where a measure of water loss rate is not required, water vapor is best added back into the air stream after CO_2 is scrubbed. This needs to be done carefully because CO_2 dissolves in water, so it is difficult to rehumidify air without adding CO_2 (see Chapter 4).

In all cases, care should be taken to keep the incurrent gas concentrations as stable as possible. One good method for achieving this is to pump the incurrent air from outside the building or, if this is not practical, from a large carboy that will act as a low-pass filter for fluctuations in room concentration levels (see Figure 4.2). Using both approaches is ideal. Alternatively, a cylinder of compressed air can be used. As discussed in Chapter 8, chamber construction is important. Every effort should be made to prevent a straight path from the incurrent to excurrent gas ports, or pronounced instability in readings caused by poor mixing will result. Apart from this caveat, the volume of the chamber should be minimized or the time constant of the system may become excessive. The alternative method of reducing the time constant (increasing the flow rate) is more problematic. Because it reduces the deflection of gas concentrations caused by the specimen, it increases the relative contribution of noise and drift from the gas analyzers.

Measuring Metabolic Rates: A Manual for Scientists. Second Edition. John R. B. Lighton, Oxford University Press (2019).
© John R. B. Lighton (2019). DOI: 10.1093/oso/9780198830399.001.0001

Almost all incurrent flow measurement systems are of the push type and are usually referred to as operating in *push mode*. In these systems, leaks downstream from the chamber are not quite as significant as in a pull system, but leaks in the chamber itself are a serious matter. Thus, the chamber (and everything from the flow meter to the point of gas analysis, or the point from which the air is subsampled, as applicable) should be carefully tested for leaks. For the chamber, sealing off both ends of the system, pressurizing it slightly, and measuring the internal pressure of the system over a period of a few seconds will be adequate. A water or oil manometer, a sensitive digital pressure gauge, or a low-pressure meter (with a pressure range of up to a few centimeters of water column) will work well.

Baselining is an extremely important aspect of flow-through respirometry. I doubt that it's possible to exaggerate the importance of baselining. As explained in Chapters 8 and 9, baselining is simply the act of measuring the incurrent concentration of the gases in the respirometry system. You can do this manually or automatically; which technique you use is optional, but baselining itself is compulsory and should be frequent. Just how frequent depends on the magnitude of analyzer or F_i drift relative to the F_i versus F_e signal you're getting from the organism. Baselining intervals may be as long as an hour or two and can be as short as a few minutes. In all of the flow-through respirometry diagrams that follow in this and subsequent chapters, I include suggestions for baselining and for the analysis of baselined data.

OXYGEN ONLY

Figure 10.1 shows a very simple system for measuring VO_2 only. Figure 10.2 shows its equivalent using a differential or dual-absolute O_2 analyzer. If both water vapor and CO_2 are scrubbed from the excurrent air stream, the correct equation is:

$$VO_2 = FR_i \left(F_iO_2 - F''_eO_2 \right) / \left(1 - F''_eO_2 \right) \qquad (10.1)$$

where FR_i is the incurrent mass flow rate scrubbed of water vapor and CO_2 (see Chapter 9, where the other variables are also defined). VO_2 takes on the same units as FR_i. Note that this equation is not dependent on VCO_2 and is thus independent of RQ.

It's worth spending a moment considering how to deal with the F''_eO_2 term. Normally, you are interested only in the $(F_iO_2 - F''_eO_2)$ term, and this is usually processed from the raw O_2 data by subtracting out baselines (F_iO_2), with drift correction as necessary, and then flipping the trace (multiplying it by –1) to obtain $(F_iO_2 - F''_eO_2)$ (Figure 10.3). How do you reconstitute F''_eO_2 for use in the denominator of the above equation? The easiest way is to copy the original, unbaselined data to another channel before baselining it. Alternatively, we normally assume that F_iO_2 is 0.2094, the fractional O_2 concentration in dry, CO_2-free air. Thus, another way to reconstitute F''_eO_2 is to create an additional channel during data analysis in which the $(F_iO_2 - F''_eO_2)$ term, obtained as described above, is

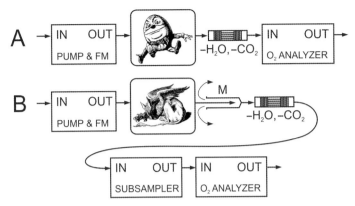

Figure 10.1 A simple push mode O_2 respirometry system (incurrent flow measurement). (A) No subsampling, typical flow rate < 1 L min^{-1}. Suitable for animals as large as rats or smaller. (B) With subsampling with manifold (M), typical main flow rate 1 L min^{-1} and more (larger animals), subsample flow rate typically150–200 mL min^{-1}. Scrubber chemicals may not be required, depending on the goals of your experiment (see Chapters 8 and 9).

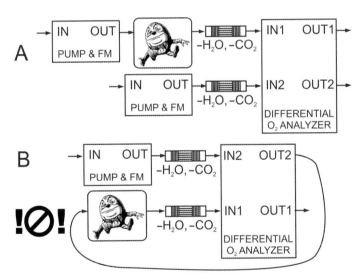

Figure 10.2 A push mode O_2 respirometry system using a differential or dual-absolute O_2 analyzer. Shown without subsampling. (A) Preferred method that maintains minimal error from fluctuations in F_iO_2 and equalizes flow, and thus pressure, between the differential analyzer's channels. (B) Less satisfactory but simpler method that imposes pressure differences between analyzer channels, adding to noise and drift. Subsampling requires two manifolds and subsamplers for option A and one of each for option B. Scrubber chemicals may not be required, depending on the goals of your experiment (see Chapters 8 and 9).

subtracted from 0.2094. Either approach is easy to implement with a good data analysis system, or even with a spreadsheet. In the event that $(F_iO_2 - F''_eO_2)$ is small (less than, say, 0.001, or 0.1 percent), you would not appreciably impact the error budget of your system by simply using $1 - 0.209 = 0.791$ as the denominator.

If you are not scrubbing CO_2 (and you are not measuring it) but you are scrubbing water vapor, then the appropriate equation is:

$$VO_2 = FR_i \left(F_iO_2 - F'_eO_2 \right) / \left[1 - F'_eO_2 \left(1 - RQ \right) \right] \tag{10.2}$$

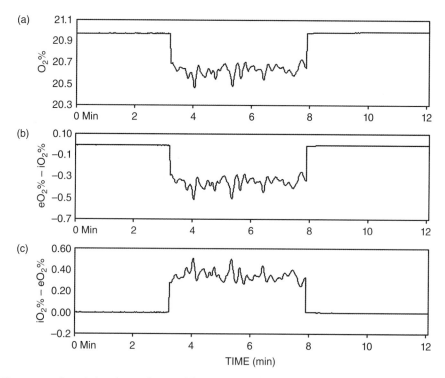

Figure 10.3 Steps in baselining the signal from an O_2 analyzer. The signal is from a small mammal at a flow rate of 1000 mL min^{-1} (Lighton, unpublished data). The recording begins and ends with baseline (incurrent) values of O_2 concentration, expressed in percent. In the middle the recording was paused and the excurrent air from the respirometer chamber was sampled. Then the recording was paused again and baseline air was sampled again. This procedure can be automated for one or multiple chambers. (A) Raw output of the O_2 analyzer in percent O_2. (B) Baseline is corrected (i.e., corrected for drift from the beginning to the end of the recording, assuming a linear drift, rotated, and subtracted). This is equivalent to subtracting incurrent O_2 percent from excurrent O_2 percent, while simultaneously correcting for any drift. (C) The trace is flipped (multiplied by –1), yielding incurrent O_2 minus excurrent O_2, as required by the equations (Chapter 9). Before transforming the data with an appropriate respirometry equation (e.g., equation 10.1), the trace would be divided by 100 to yield $(F_iO_2 - F_eO_2)$.

Note the RQ term, which is the respiratory quotient, or VCO_2/VO_2. This is annoying, but there's no avoiding the fact that F_eCO_2 is now diluting F_eO_2, and thus its concentration must be known or guessed at. Since you are not measuring CO_2, you have to guess at it, and if you assume a consensus RQ of 0.8 (unless you have a good reason to guess otherwise), the error will be relatively small, but nevertheless appreciable; about −3 to +5 percent, as the actual RQ varies from 0.7 (pure lipid catabolism) to 1.0 (pure carbohydrate catabolism). Note that if you are primarily interested in energy expenditure in Watts, the oxyjoule equivalent of O_2 (from which gas exchange data are translated into energy equivalents) also varies with RQ, and the direction of this variation very nearly cancels the effect of RQ on VO_2 (Weir, 1949; Koteja, 1996; see the end of Chapter 9 for more).

CARBON DIOXIDE ONLY

Carbon dioxide only is the most sensitive form of flow-through respirometry. With a high-end CO_2 analyzer and scrupulous baselining, it is easily possible to get accurate results from single small insects weighing < 0.001 g, such as the important model organism *Drosophila* (see, e.g., Djawdan et al., 1997; Lighton, 2007; Lighton and Schilman, 2007). A typical push mode setup, where flow is measured on the incurrent side, is similar to that shown in Figure 10.1, except that a CO_2 analyzer rather than an O_2 analyzer is used and no CO_2 absorbent is used.

Remember that if you choose to scrub water vapor from the excurrent air, you should use Drierite only if you are willing to put up with extremely sluggish response times. You should not use silica gel at all. The only chemical scrubber that will give good results is magnesium perchlorate; alternative scrubbers include condensers and chemical-free, selectively permeable membrane-based scrubbing systems (Chapter 19).

If you are not measuring O_2 but are only measuring CO_2, congratulations; your computations may have become enormously simpler. This is especially true if you are measuring a very small organism that does not consume enough O_2 to significantly reduce F_eO_2 (thus concentrating F_eCO_2) and that does not add a significant F_eH_2O to the excurrent air (thus diluting F_eCO_2). In such a case, you can eliminate the water vapor scrubber and use this simple equation:

$$VCO_2 = FR_i \left(F_eCO_2 - F_iCO_2 \right) \tag{10.3}$$

This depends, of course, on low cross-sensitivity of the CO_2 analyzer to water vapor. Most CO_2 analyzers are oblivious to water vapor because it absorbs infrared radiation outside the 4.26-μm CO_2 band (see Chapter 16 for more details), but you should test yours to be sure. As a general rule, if F_eO_2 is within 0.001 of F_iO_2 and F_eH_2O is within 0.001 of F_iH_2O, using this simplifying equation will not contribute significantly to the system's error budget.

This is not an acceptable shortcut for larger organisms, however. It will work for an ant but should be eschewed for a mouse, shrew, or finch, let alone anything larger. The following equation assumes that you are scrubbing water vapor, and you should do so (or allow for it as explained above) now that it has become significant.

$$VCO_2 = FR_i \left(F'_e CO_2 - F_i CO_2 \right) / \left\{ 1 - F'_e CO_2 \left[1 - \left(1 / RQ \right) \right] \right\} \tag{10.4}$$

Note that the magnitude of the RQ correction is dependent on the magnitude of $F'_e CO_2$, which makes sense because it's a concentration correction; the more CO_2 produced, the more O_2 must be consumed, which will concentrate the CO_2 by an amount depending on the RQ, the ratio of CO_2 produced to O_2 consumed. If $F'_e CO_2$ is tiny, the correction term will be too, relative to other sources of error in the system, notably flow rate measurement (Chapter 17).

If you want to eliminate water vapor scrubbing and are able to measure water vapor pressure and barometric pressure (some good-quality CO_2 analyzers have barometric pressure meters built in), see the section on "Oxygen only" measurements for guidance. Also, be sure to test for cross-sensitivity between CO_2 and water vapor.

BOTH OXYGEN AND CARBON DIOXIDE

Measuring both gas species gives the most accurate results and allows you to mathematically compensate for the enrichment of CO_2 caused by O_2 consumption and the O_2 dilution caused by CO_2 production on a moment-by-moment basis. Measuring both O_2 and CO_2 also gives you access to the organism's RQ, which can be extremely informative. Recent developments in top-of-the-line O_2 analyzers even allow RQ determination in individual small insects in real time, facilitating the development of insect models for clinical conditions such as metabolic syndrome (Schilder and Marden, 2006). It's worth mentioning here that RQ may, on occasion, stray beyond the bounds of 0.7–1.0. It can exceed 1 if an animal is synthesizing fat (*de novo* lipogenesis), and it can be artificially displaced by changes in CO_2 solubility. For example, this sometimes happens over periods of many hours when body temperature fluctuates in large ectotherms (Chappell, personal communication). CO_2 can also be released in large quantities, driving the measured RQ above 1, in the presence of anaerobiosis and resulting lactate accumulation.

If you need to measure both O_2 and CO_2, it's usually important to ensure that the O_2 and CO_2 signals are in phase—that is, not lagged in time with respect to one another. The same holds true for the water vapor signal if water vapor is not scrubbed from the excurrent air stream. It will certainly be necessary to shift the recorded channels of data in time so that they line up. The first analyzer to encounter excurrent air is the leader analyzer; the others are followers. The data from the followers will need to be moved leftward or back in time until they match the data from the leader. This is easy to do with a spreadsheet or a good data analysis program. It is, of course, necessary to determine how many samples to move each follower leftward. This is usually determined by introducing

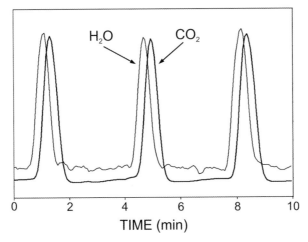

Figure 10.4 Correcting for lag between different analyzers in a respirometry system. Data show discontinuous gas exchange from the ant *Camponotus vicinus* (Lighton and Turner, unpublished data) using push mode CO_2 respirometry. Y-axis scaling is not shown. The water vapor analyzer (thin trace) is first in the analyzer chain, and so it reacts first to the excurrent signal. This is easily corrected by sliding the lagging signal (CO_2) leftward in the data analysis software.

a short disturbance into the system (such as blowing into an open syringe and then injecting the air into the sample air stream) while recording the result. The follower lags can then be determined by visual inspection or cross-correlation analysis. Figure 10.4 shows lagged data, which are easily corrected by good software. To aid in determining lag times, it may be necessary to use the instantaneous correction (Chapter 8) judiciously to make the analyzers' response curves resemble each other as closely as possible. Of course, this is much less of a problem if the data change relatively slowly compared to the response speeds of the gas analyzers, which is usually the case in flow-through respirometry.

Of the various possible ways to inter-correct O_2 and CO_2 data, the most straightforward is to use the various fractional concentrations to account for dilution and concentration effects. Thus, for dry air,

$$VO_2 = FR_i \left[\left(F_i O_2 - F'_e O_2 \right) - F'_e O_2 \left(F'_e CO_2 - F_i CO_2 \right) \right] / \left(1 - F'_e O_2 \right) \qquad (10.5)$$

and

$$VCO_2 = FR_i \left[\left(F'_e CO_2 - F'_i CO_2 \right) - F'_e CO_2 \left(F_i O_2 - F'_e O_2 \right) \right] / \left(1 - F'_e CO_2 \right) \qquad (10.6)$$

This set of two equations is critically dependent on accurate response and phase matching between the O_2 and CO_2 signals.

There are, of course, other ways to do this. One is to take advantage of CO_2-scrubbed excurrent air and an incurrent flow meter and use equation 10.1 for measuring VO_2, which in this form is independent of VCO_2. Then we can use the calculated VO_2 to go back and calculate VCO_2. This obviously requires that the CO_2 analyzer be the leader, followed by a CO_2 scrubber and then the O_2 analyzer. Armed with our VCO_2-independent VO_2, we can calculate VCO_2 thus:

$$VCO_2 = FR_i \left[\left(F'_e CO_2 - F_i CO_2 \right) - F_e CO_2 \left(VO_2 \right) \right] / \left(1 - F_e CO_2 \right) \qquad (10.7)$$

EVAPORATIVE WATER LOSS

As can be shown from the equations presented in Chapter 9,

$$VH_2O = FR_i \left(F_e H_2O - F_i H_2O \right) / (1 - F_e H_2O) \qquad (10.8)$$

Recall that at STP, each milliliter of water vapor is equivalent to 0.803 mg H_2O. In practice the $(1 - F_e H_2O)$ term generally has no significant effect and can usually be ignored.

Alternatively, you can measure water vapor partial pressure and ambient temperature and convert the two into water vapor density thus:

$$WVD = WVP / \left(T \times R_w \right) \qquad (10.9)$$

where WVD is water vapor density in micrograms per milliliter, WVP is water vapor pressure in kilopascals, T is temperature in Kelvin ($°C + 273.15$), and R_w is the gas constant for water vapor, which is 461.5 J kg^{-1} K^{-1}. Thus armed with WVD, multiply it by FR$_i$ in the appropriate units to yield water loss rate in gravimetric units. For example, multiplying WVD by FR$_i$ in milliliters per minute yields water loss rate in micrograms per minute; multiplying the result by 0.06 yields units of milligrams per hour.

For completeness, I mention an archaic but effective technique for determining evaporative water loss rate. If there is a water vapor scrubber column in the system, simply weighing it before and after a session of flow-through respirometry will give you the mass of water vapor lost by the animal during that time period. If you are paranoid (which is a useful attribute in research), you can opt to measure water loss by both techniques. In an early validation study, Bernstein et al. (1977) used both flow-through and desiccator-gravimetric water loss measurements and found that the two agreed exactly within experimental error. In their case they determined the water vapor partial pressure of the excurrent air stream from its dewpoint, using the empirical equation

$$WVP = 0.95T / \left(T + 265.5 \right) + 0.7858 \qquad (10.10)$$

where WVP is in kilopascals and T is the measured dewpoint. I mention this because dewpoint analyzers lurk at the back of closets, unused, in some laboratories or can be purchased inexpensively through on-line auctions. If their calibration is checked by their manufacturer, they are capable of very accurate water vapor analysis (Chapter 16).

EXTENDING TO MULTIPLE CHAMBERS

Here we come to a parting of the ways between systems for insects and other small, hypoxia-tolerant animals and systems for most vertebrates, especially mammals and birds. However, regardless of specimen size, all multiple-chamber systems use a *multi-plexer*, a device that directs a stream of air along a chosen path under computer control. Multiplexers are described in some detail in Chapter 4; as discussed in that chapter, they can be used for closed-system (or stop-flow) respirometry as well.

As with single animals, activity detection is important to provide information about whether the animal's metabolic state is contaminated by activity (Chapter 18). Different activity levels can also be interesting in their own right. Multiple activity detectors can monitor multiple animals, and only the signal from the activity detector associated with the selected chamber needs to be monitored. Good-quality multiplexers can electrically select a given activity detector's analog signal and steer it to a single channel in the data acquisition system, at the same time as directing actual air flow through the chamber associated with that activity detector.

Small animals

For small animals, the respirometer chambers are usually small as well, and plumbed so that only one chamber at a time can be selected for air to pass through it (Figure 10.5). Typically, the main flow rate and the subsampling flow rate are equivalent; in other words, the entire main flow is pushed through the chambers and the analyzers. In implementations involving insects, this flow rate is typically 50–200 mL min^{-1}. The chambers that are not selected can, depending on the configuration of the multiplexer, be left

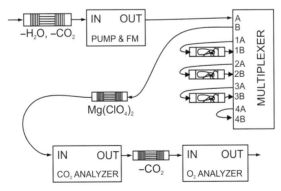

Figure 10.5 A simple multiple-chamber, push mode respirometry system for small animals. Three animals plus baseline (no chamber) are shown. This system, as pictured, does not flow air through chambers that are not selected, causing O_2 depletion and CO_2 buildup unless a chamber is selected. This can be mitigated using the method shown in figure 10.7. Scrubber chemicals may not be required, depending on the goals of your experiment (see Chapters 8 and 9).

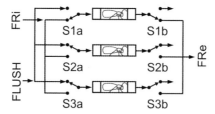

Figure 10.6 Setup as in figure 10.6, but arranged so that a low flow rate is flushed through nonselected chambers. The changeover solenoid gas switches are shown in conceptual form. Three channels are shown. Only one channel is selected at a time. As shown, the first channel is selected, and the incurrent flow rate, FR_i, passes only through that channel's chamber, heading for the analyzers as excurrent flow, FR_e. The other, nonselected chambers are connected to air stream FLUSH, which optionally can be derived from the nearly dry, nearly CO_2-free flow leaving the analyzers. Plumbing the system in this way allows for much faster switching between chambers because it is not necessary to include a long washout period before measurement, especially if FR_i and the per-chamber flush flow rates are matched. Even if the flow rates are not closely matched, the improvement is considerable.

sealed or, more optimally, can have a low rate of flow pass through them (Figure 10.6)—ideally the same flow rate as during measurements.

The data acquisition system is programmed to switch between the chambers while recording. A marker, if available in the data acquisition system, should be placed at the point of each switchover. At the point of switchover, a bolus of O_2-depleted and CO_2-enriched air will sweep through the analyzers if the chambers were completely closed while they were not selected. This effect, reminiscent of constant volume analysis (Chapter 4), is much less significant if flow occurred during the time the chambers were not selected. Because the bolus will be ignored by most researchers, it is common practice to pause the recording until readings have stabilized—typically a minute or so. Then the recording is taken off pause, and the animal's F_e readings are taken. After an interval that depends on the object of the research but can be anywhere from minutes to hours, the multiplexer is commanded to switch to another chamber, or to an empty chamber that serves as the system's baseline (that is, its method of determining F_i values). Another marker celebrates the occasion. Some workers alternate baseline and sample readings; others take one baseline at the beginning of a recording and at the end, with animal chambers being switched in between. As always, a tradeoff is involved. Taking a baseline between each chamber increases the accuracy of the measurements but decreases the "duty factor" (percent of time monitored) for each animal.

Workers who take no baselines are foolish and will not be considered further in this book. Those who take too few baselines are not necessarily foolish but are unlikely to be awarded tenure.

With appropriate software and properly coded markers, it is easy to compensate for analyzer drift, obtain the $F_e - F_i$ values, and calculate VO_2 and/or VCO_2 (depending on which analyzers are used) for each chamber. At this point, the further data analysis strategies diverge, depending on the aims of the researcher.

Commonly, a mean value of gas exchange is determined between successive markers, and the result is written on a spreadsheet, with each chamber in a separate column and with successive recordings (each of which scans through all chambers, plus baseline) in successive rows. Alternatively, a low or level section of a specified length is found, and its mean is taken and written as above (see Chapter 15 for more on data selection). It is also possible, of course, to program software to do all of these operations behind the scenes without an opportunity for user interaction and without recording each scan through the chambers. However, if this path is chosen, faults or problems may not become evident until it is far too late to save a large body of data, and the complete raw data will not be available for a rescue effort or for leaving an audit trail. It is best to record data in the most basic form and then process it through analysis, which with suitable software can be completely automated. In my opinion and experience, the completely automated approach without raw data storage, although appealing, should be avoided at all costs. Chapter 13 goes into more detail on this topic.

Large animals

For large animals, the technique described above will not work well. Each animal inhabits a fairly large chamber through which a fairly high rate of flow is pushed. If that flow stops, the animal may become asphyxiated and die within a matter of minutes. Even if the animal survives, it will be many minutes before equilibrium gas values are attained, and the animal will have been at worst highly stressed and at best aware that an aspect of its environment suddenly changed and is thus unlikely to reward you with good data.

Several solutions exist to address this problem. All of them amount to different ways of fulfilling the requirement for permanent flow through each chamber. The approaches differ chiefly in cost, accuracy, and ease of use, with higher cost bringing greater accuracy and greater ease of use, as is usually the case in research.

For best results, whichever configuration described below is chosen, one incurrent flow meter per chamber will be required, with the possible exception of the mass flow control option. At flow rates up to about 1 L min^{-1}, it is possible (as shown in Chapter 11) to share one *excurrent* mass flow meter between all chambers; sharing a meter at higher flow rates is not recommended. With incurrent flow measurement, you have little choice but to use independent mass flow meters.

Measuring multiple large animals such as mammals or birds is a subject unto itself, and is covered in depth in Chapter 13. However, some more basic approaches are described below.

The low-cost approach

The low-cost approach is to use one pump (usually a substantial diaphragm pump) and a nest of needle valves, one per chamber, to drive flow through each chamber (Figure 10.7). Multiple pumps and needle valve nests can be used. Select good-quality needle valves (see Appendix 1). Ideally, use one flow meter per valve and adjust each valve to give the

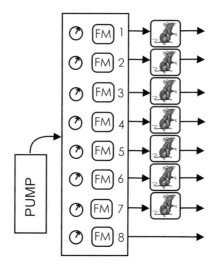

Figure 10.7 A simple approach for maintaining flow through multiple respirometer chambers—required for animals with high metabolic rates, such as mammals. A single pump supplies an array of needle valves for manual flow control. On the output of each needle valve, a mass flow meter measures the flow rate into each chamber.

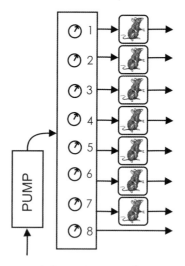

Figure 10.8 The lowest cost approach for pushing continuous flow through multiple chambers. A single pump supplies an array of needle valves for manual flow control. Flow rates are adjusted to the desired value during setup. No individual flow metering takes place.

same flow rate as the others. Instruments exist that are designed for this purpose and incorporate multiple, calibrated mass flow meters (see Appendix 1). They can also eavesdrop on the commands sent to the multiplexer and direct only the selected mass flow rate data to the data acquisition system. Although the flow rates may be only approximately equal, this is not a problem because all incurrent flow rates are measured, and the flow rate into the selected chamber, as measured by the data acquisition system, is used for the respirometry calculations.

If low cost is all important, you can use the equal-flows, meterless approach (Figure 10.8). Adjust the valves until the flow through all channels is identical (as measured with one flow meter that you move from channel to channel). This will require great patience because it is an iterative process; each adjustment you make will affect the other channels, though this effect declines as flow rates progressively equalize. You will then use the measured and set flow rate for all chambers during data analysis and pray that it is accurate.

The equal-flows, meterless approach

The equal-flows, meterless approach relies on the dubious assumptions that (1) the flow rates are, indeed, equal; (2) they will stay equal; (3) they will maintain the same value as initially measured; and (4) the insertion of the flow meter doesn't affect resistance, and hence flow, through the chamber it's monitoring. To some extent you can control for the third assumption by keeping a single mass flow meter in the incurrent flow of one chamber.

The easiest approach

The easiest approach is to assign a mass flow control valve to provide the incurrent flow to each chamber (Figure 10.9). The high-pressure air supply for each valve can be derived from a common pump or from a compressed air source, or each mass flow control can be assigned its own pump. A conceptually similar approach is to use a mass flow meter and a pump in a closed-loop control system to generate the incurrent air flow for each chamber. Either way, the cost of such a flow-generating system is quite high; about US\$2000–3000 per channel. The individual flow rates can be electrically selected by the multiplexer control signal, in some cases using a subsidiary electrical multiplexer, for use

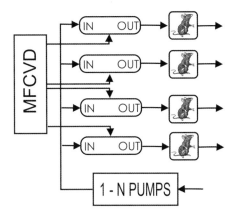

Figure 10.9 The easiest (but highest cost) approach for pushing continuous flow through multiple respirometer chambers. Individual mass flow control valves regulate a fixed flow rate into each chamber (only four chambers shown). They can be driven from individual pumps or a high-capacity pump as shown. Mass flow control valve driver units (MFCVD) can typically control one, two, or four mass flow control valves, depending on make and model.

during data analysis. However, many workers simply assume that the set flow rate is constant and use a fixed value for flow rate during data analysis. They will usually be correct.

DIRECT SAMPLING OR SUBSAMPLING?

I assume that you have plumbed your chambers so that a known or measurable flow rate is entering each chamber. How do you decide how to plumb the outlets of the chambers? You need to make the choice between directing the flow from all chambers into the multiplexer, with one selected flow being directed to the gas analyzers, and directing the flows from the chambers into manifolds, from which one selected flow is subsampled and pulled through the gas analyzers.

If the main flow rate through each chamber is less than about 1 L min^{-1}, it is easiest to direct all of the flows through the multiplexer. Good-quality multiplexers can be set to allow free gas flow from nonselected chambers, which is obviously important in this application. Optionally, you can insert a flow meter in the incurrent air line, thus measuring the incurrent flow rate being pushed through the selected channel (Figure 10.10). After each chamber, it is wise to place a filter of some kind, to prevent hair, dander, dust, and so on from the chamber's occupant from collecting in the multiplexer's solenoid

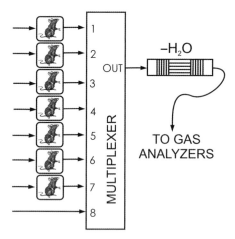

Figure 10.10 Selecting excurrent flow from multiple chambers using direct connection of the excurrent flows to a multiplexer. This is generally recommended only for flow rates < 1 L min^{-1}. In this case (depending on the flow capacities of the analyzers), a separate subsampling flow system is not required. However, a multiplexer that vents nonselected flows is required in this application. A water vapor scrubber is shown; this would typically be replaced with a water vapor and CO_2 scrubber for an O_2-only system, or a water vapor analyzer for water vapor dilution correction in conjunction with barometric pressure measurement (equation 8.7). Optionally, a mass flow meter can be placed after the multiplexer's output to measure the excurrent flow rate of the selected chamber.

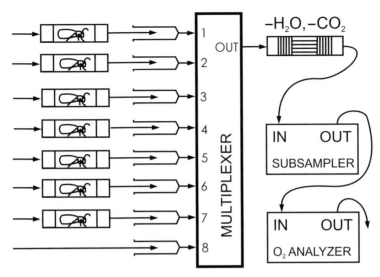

Figure 10.11 Selecting excurrent flow from multiple chambers using an indirect connection of the excurrent flows to a multiplexer, via manifolds. This approach is generally necessary if chamber flow rates exceed about 1 L min^{-1}. A subsampling system is required. An O_2-only measurement system is shown. Scrubber chemicals may not be required, depending on the goals of your experiment (see Chapters 8 and 9).

valves, which will eventually affect its performance. A small syringe barrel filled with fiberglass wool will work adequately.

For higher flow rates, subsampling is required. Each chamber's outlet empties into a manifold, which is simply a place from which a subsample of the air can be pulled. A section of large-diameter tubing, an empty syringe barrel, a short section of PVC pipe, or a suitably configured T fitting—any of these can be successful manifolds; use your imagination. From these manifolds, one sample is selected with the multiplexer and pulled through the gas analysis chain (Figure 10.11). It is important to keep the subsampled air flow to (as a rule) a third or less of the main flow rate. About 200 mL min^{-1} is a common value.

The subsampled flow rate does not enter into any of the respirometry calculations; only the main flow through each chamber is significant. It is, however, good practice to record the subsampled air flow when practical.

See Chapter 13 for more details on measuring metabolic rates of multiple animals.

11

Flow-through Respirometry: Excurrent Flow Measurement

Much of this chapter assumes knowledge of points raised in Chapter 10, on topics such as organizing flow, chamber design, and so on. If you have not already read that chapter, this would be a good time to do so. A lot of material relevant to both types of respirometry was covered there and will not be repeated here.

Most systems that measure excurrent rather than incurrent flow—in other words, with the measured flow rate emerging from rather than entering into the system being monitored—are of the unsealed kind, though this does not have to be the case. They are usually based on pulling the air through the system rather than pushing air through it, as discussed in Chapter 10. Most such systems use ambient air at the entry point of the chamber as the baseline. This is always the case with masks, for example. These characteristics often make these systems simpler and more practical to set up than push systems with incurrent flow rate measurement, especially for larger animals or when the intention is to measure metabolic rate during intense activity, such as treadmill running.

However, as discussed in Chapter 9, excurrent flow rates are made up of incurrent flow rates plus water vapor production by evaporation, plus CO_2 production, minus O_2 consumption, all of which need to be accounted for. Fortunately, baselining a system based on measurement of excurrent flow rate is quite simple and usually consists of simply diverting the subsampling system that pulls air through the gas analyzers so that it measures ambient air rather than the excurrent air. F_iO_2, F_iCO_2, and F_iH_2O are all easily measured, as are F_eO_2, F_eCO_2, and F_eH_2O. Thus armed, we can proceed.

Figure 11.1 shows a typical single-chamber system or mask with excurrent flow rate measurement (multiple-chamber systems are covered at the end of this chapter). The subsequent equations assume that you are drying the air leaving the chamber before measuring its flow rate; if that is not the case, you will need to correct the flow rate for the presence of water vapor as shown in equation 8.6.

OXYGEN ONLY

If you are drying and removing CO_2 from the air prior to measuring its flow rate and passing it (or a subsample) through the O_2 analyzer, the following equation holds:

Measuring Metabolic Rates: A Manual for Scientists. Second Edition. John R. B. Lighton, Oxford University Press (2019).
© John R. B. Lighton (2019). DOI: 10.1093/oso/9780198830399.001.0001

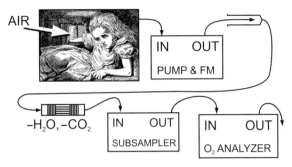

Figure 11.1 A simple pull mode O_2 respirometry system using excurrent flow measurement. This system uses a mask (shown here, Alice is inside the rather cramped rabbit's house that functions as a mask). Air enters a mask through any available spaces, in this case an open window. Baselining consists of recording from the mask without the animal present or disconnecting the mask and sampling the ambient air. Note that the excurrent flow includes more water vapor and CO_2 and less O_2 than the incurrent flow (see Figure 9.1). The system as shown does not compensate for these effects on excurrent flow rate, especially addition of water vapor. One solution is to scrub water vapor from the main flow before measuring its flow rate. This is often impractical; see equation 8.6 for an alternative to drying the entire primary flow. Scrubber chemicals may not be required, depending on the goals of your experiment (see Chapters 8 and 9).

$$VO_2 = FR_e \left(F_i O_2 - F''_e O_2 \right) / \left(1 - F_i O_2 \right) \tag{11.1}$$

where the terms are as defined in Chapter 9. Note that this is a rather friendly equation, in that (1) we do not need to keep track of $F_e O_2$ so that we can recycle it in the denominator (as in the incurrent flow measurement case in Chapter 10), and (2) it is independent of RQ. $F_i O_2$ is usually, by definition, 0.2094 for air scrubbed of water vapor and CO_2 and can be treated as a constant unless you have good reason to know that it takes a different value.

The $(F_i O_2 - F'_e O_2)$ term is usually created during data analysis by subtracting the baseline ($= F_i O_2$ by definition) values, preferably with drift correction (Chapter 10), and then flipping the trace (i.e., multiplying it by –1) to invert the order of subtraction.

If you are only drying the air entering the O_2 analyzer and flow meter and not scrubbing CO_2 from it at any point, including prior to O_2 analysis, then this equation holds:

$$VO_2 = FR_e \left(F_i O_2 - F'_e O_2 \right) / \left[1 - F_i O_2 \left(1 - RQ \right) \right] \tag{11.2}$$

Again, note that the denominator uses $F_i O_2$, which can generally be treated as a constant of 0.2094. Of course, CO_2 is now diluting O_2, and the RQ term seeks to compensate for this. If an RQ of 0.8 is assumed, the error can be nearly 3 percent if the RQ is in fact 0.7, and almost 5 percent if it's 1.0 (Koteja, 1996). However, as Koteja points out, if you want to calculate energy equivalents, you can take advantage of the fact that the dependence of the oxyjoule conversion factor on RQ almost exactly counteracts this error, giving you a maximum error of only about 0.6 percent; see Chapter 10 and especially the end of Chapter 9 for more details.

If the flow leaving the chamber is measured after drying the air (or otherwise compensating for its water vapor content; equation 8.7) but before scrubbing CO_2 from it, and CO_2 is then scrubbed from the air before measuring O_2, the equation is a little more complex:

$$VO_2 = FR_e \left(F_iO_2 - F''_eO_2 \right) / \left[1 - F_iO_2 + RQ \left(F_iO_2 - F''_eO_2 \right) \right] \qquad (11.3)$$

CARBON DIOXIDE ONLY

If CO_2 is not scrubbed from the excurrent air stream before you measure the flow rate, as in a typical mask setup,

$$VCO_2 = FR_e \left(F'_eCO_2 - F_iCO_2 \right) / \left\{ 1 - F_iCO_2 \left[1 - \left(1 / RQ \right) \right] \right\} \qquad (11.4)$$

The problematic RQ term derives from the fact that uptake of O_2 concentrates the CO_2 in the sample passing through the analyzers and reduces the excurrent flow rate. However, notice that the RQ term drops out if F_iCO_2 is equal to zero. In other words, if you scrub CO_2 from the incurrent air stream, you can legitimately use the old standby, equation 10.3:

$$VCO_2 = FR_e \left(F'_eCO_2 - F_iCO_2 \right)$$

Even when you don't scrub incurrent CO_2, as in a typical mask system, the error from using the simpler equation in typical situations where F_iCO_2 is about 0.05 percent is well under 0.1 percent. It's nearly impossible to measure air flow with 0.1 percent accuracy, whatever some vendors may tell you, so worrying about that error is akin to tilting at windmills, which is not to say you shouldn't do it if you find it to be exciting.

If CO_2 is scrubbed from the air stream before flow measurement, which would generally require the entire unscrubbed flow rate to pass through the CO_2 analyzer and is therefore an unlikely configuration, you should consider this:

$$VCO_2 = FR_e \left(F'_eCO_2 - F_iCO_2 \right) / \left(1 - F'_eCO_2 + F_iCO_2 / RQ \right) \qquad (11.5)$$

This equation, which compensates for the reduction in measured flow rate caused by the absence of CO_2, will start to differ significantly from the simpler equation 10.3 in terms of the system's error budget when F'_eCO_2 exceeds about 0.5 percent.

BOTH OXYGEN AND CARBON DIOXIDE

Measuring both O_2 and CO_2 opens many possibilities, depending on the exact configuration of analyzers, scrubber, and flow meter. Usually the flow meter will come first, and the measured flow rate will be corrected for the presence of water vapor (equation 8.6).

Then a subsample of the air stream will be taken, and water vapor and possibly (after the CO_2 analyzer) CO_2 will be scrubbed from it. However, as you may have gathered from the preceding discussions, VO_2 and VCO_2 interact strongly during pull system data analysis. Therefore, you will usually need to use the fractional concentrations of both O_2 and CO_2 to calculate either VO_2 or VCO_2.

If you scrub CO_2 prior to both flow measurement and O_2 analysis, however, you can take advantage of the RQ independence of VO_2 calculated as in equation 11.1. Thus armed with an accurate and RQ-independent value of VO_2, you can calculate VCO_2, taking into account the enrichment of CO_2 caused by O_2 consumption:

$$VCO_2 = \left[FR_e \left(F'_e CO_2 - F_i CO_2 \right) - F_e CO_2 \left(VO_2 \right) \right] / \left(1 - F_e CO_2 \right) \tag{11.6}$$

This approach is not practical at higher flow rates because scrubber use will become excessive. Rather, the better approach with high flow rate pull systems is not to scrub CO_2 from either the incurrent or the excurrent air streams and, where practical, to "dry" the measured flow rate by measuring water vapor pressure and applying equation 8.6. The subsampled air stream can then be dried and analyzed. The fractional concentrations of O_2 and CO_2 must be in phase during data analysis. Thus, the respective lag times of the analyzers must be corrected for and their response characteristics made as similar as possible as discussed in Chapter 8, prior to final transformation into VO_2 and VCO_2. This is especially critical for rapidly changing data. Alternatively or in addition, fractional concentrations can be averaged over periods long enough to make phase difference or response time effects insignificant.

Considering the case where CO_2 is not scrubbed from the air stream at any point, and expressing O_2 and CO_2 data in fractional concentrations only,

$$VO_2 = FR_e[(F_iO_2 - F'_eO_2) - F_iO_2(F'_eCO_2 - F_iCO_2)]/(1 - F_iO_2) \tag{11.7}$$

and

$$VCO_2 = FR_e \left[\left(F'_e CO_2 - F_i CO_2 \right) + F_i CO_2 \left(F_i O_2 - F'_e O_2 \right) \right] / \left(1 + F_i CO_2 \right) \tag{11.8}$$

Like the similar examples in Chapter 10, this set of two equations is critically dependent on accurate response and phase matching between the O_2 and CO_2 signals.

EVAPORATIVE WATER LOSS

See Chapter 10 for a treatment of this topic. The correct equation for excurrent flow measurement is:

$$VH_2O = FR_e \left(F_e H_2 O - F_i H_2 O \right) / \left(1 - F_i H_2 O \right) \tag{11.9}$$

but again, the $(1 - F_iH_2O)$ divisor is generally so close to unity that it can usually be ignored.

The reader can convince himself or herself of this by comparing the different approaches, using typical values.

EXTENDING TO MULTIPLE CHAMBERS

Most of the explanations and comments in the corresponding section of Chapter 10 are also relevant to this discussion and are not repeated here. The remainder of this chapter concentrates on the differences between incurrent and excurrent flow measurement in multiple-chamber systems.

Small animals

The setup for using a single flow meter for small animals that do not require continuous air flow through their chambers is basically identical, except that the position of the flow meter is moved from the incurrent to the excurrent side of the chambers (Figure 11.2), typically after scrubbing water vapor from the excurrent air stream. If water vapor is not scrubbed before flow measurement, it should be measured and accounted for as shown in equation 8.6.

Large animals

For larger animals, continuous flow through all chambers is necessary, as described in Chapter 10.

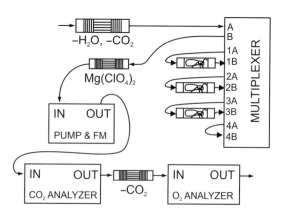

Figure 11.2 A simple multiple-chamber pull system, with a single flow meter. Water vapor is absorbed before flow measurement; alternatively, the flow rate can be corrected by measuring its water vapor content. Air flows only through the selected chamber, so this setup is only suitable for small animals such as arthropods or small mammals such as mice unless a high-flow multiplexer and a subsampling system is used. Scrubber chemicals may not be required, depending on the goals of your experiment (see Chapter 8 and 9).

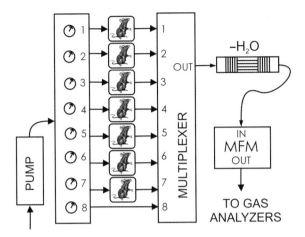

Figure 11.3 A simple multiple-chamber pull system that maintains flow through all chambers, using a pump and multiple needle valves, and measures the excurrent flow rate of the selected channel with mass flow meter (MFM). A multiplexer that vents nonselected flows is required in this application. Because flow is directed through the multiplexer, flow rates < 1 L min^{-1} are recommended. Scrubber chemicals may not be required, depending on the goals of your experiment (see Chapters 8 and 9).

I consider first a low flow rate (< 1 L min^{-1}) push system with excurrent flow measurement. This has the advantages that flow through all chambers is continuous and that only one flow meter is required for any number of chambers (Figure 11.3). Continuous flow into each chamber can be provided by a pump and multiple needle valve system, as described in Chapter 10.

Because the flow rate per chamber is less than about 1 L min^{-1}, the excurrent flows from each chamber can be directed through the multiplexer straight to the scrubber (if any) and the analyzers. The chambers should be tested for leaks first and any leaks corrected. Place the flow meter in the selected output of the multiplexer as shown in Figure 11.3. With no animals in the chambers, select the flow from successive channels in the multiplexer, adjusting the appropriate needle valves until all flows are more or less equivalent. Now you can record data as described in Chapter 10, flipping between chambers or between chambers and baseline, and the single flow meter will always faithfully track the excurrent flow rate of the selected channel. Air entering the multiplexer after passing through a chamber should be filtered.

If higher flow rates are required, or if a fully sealed respirometry chamber is either impossible or undesirable, then ambient air can be pulled into almost-sealed chambers or into masks (the two function in much the same way), with the excurrent flows being continuously maintained. In this case, multiple pumps are required in practice because the flow from each chamber or mask must be not only controlled and measured, but also kept separate from other flows. Analogously to the discussion in Chapter 10, for low flow rates (< 1 L min^{-1}) this can be implemented with a single mass flow meter by plumbing

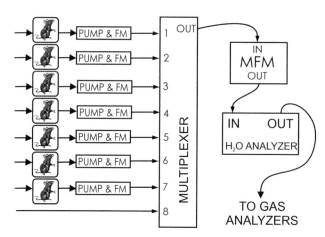

Figure 11.4 A pull system with separate pumps and flow controls on the excurrent flow side. Note that one pump (and needle valve to control flow rate) per chamber is required, but at low flow rates (< 1 L min^{-1}) the selected chamber's excurrent flow rate can be directly measured using a single mass flow meter after the multiplexer, either after drying or, if a water vapor analyzer is used, corrected for water vapor dilution (equation 8.7). A multiplexer that vents nonselected flows is required in this application.

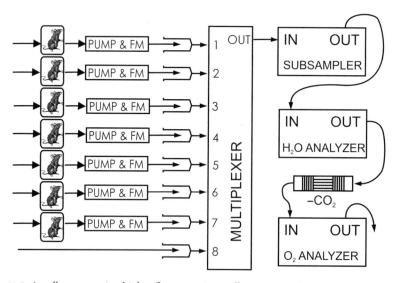

Figure 11.5 A pull system using higher flow rates (typically > 1 L min^{-1}). Note that one pump per chamber is still required, and each channel has an individual mass flow meter. The multiplexer allows a subsample from any flow to be selected via manifolds. In this case, a water vapor analyzer is shown for water vapor dilution compensation of flow rate (equation 8.6). A scrubber is shown but is not actually required if all relevant gas species (H_2O, CO_2, and O_2) are measured for comprehensive correction, for example, compensating for water vapor dilution (equation 8.7). Scrubber chemicals may not be required, depending on the goals of your experiment (see Chapters 8 and 9).

the excurrent air streams (after they have passed through the individual pumps) directly through a multiplexer (Figure 11.4).

At higher flow rates, individual pumps, needle valves, and mass flow meters for each chamber are required, with the multiplexer subsampling the analyte air stream through manifolds (Figure 11.5). An easier but more expensive approach is to use individual mass flow control valves and pumps for each chamber. If this is done, care should be taken to ensure that the mass flow control valves are specified for operating at a low differential pressure (30 kPa maximum) so that minimal strain is placed on the pumps. Alternatively, mass flow controllers that employ a mass flow meter and a controlled pump in a servo loop to maintain a fixed flow rate can be used; these are likely to have a longer service life because their pumps run only enough to maintain the set flow rate.

See Chapter 13 for more details on measuring metabolic rates of multiple animals.

The pulled excurrent air flows are then selected for subsampling and analysis identically to the methods described in Chapter 10.

A more advanced technique for implementing a multiple-channel pull flow system is described in Chapter 13.

12

Room Calorimetry and Mask Respirometry

ROOM CALORIMETRY

Powerful though it is, conventional flow-through respirometry is not a perfect technique. A bit of ingenuity will go far (see Chapter 7), but there are still situations in which using it is impractical. Among these is measuring the metabolic rate of a sizeable animal in its natural environment as it goes about its various activities. Spot measurements of metabolic rate during specific activities or at rest are useful (see section "Short-term (mask) human metabolic measurement" for more on this topic), but must necessarily give an incomplete picture. Doubly labeled water (Chapter 7) gives a good estimate of *mean* metabolic rate in terms of CO_2 emission over intervals of several hours to several days or more, but doesn't allow the researcher to dissect the energetic cost of individual activities except in the most indirect way. This is not to say that doubly labeled water has no role; merely that if a nonintegrative measure with fine temporal granularity is required, other approaches must be used.

This methodological impasse is especially frustrating for obesity studies, a field increasing in importance even faster than the expansion of the industrialized world's waistlines. Obesity studies are based on a simple theory. In an adult animal, long-term energy intake that exceeds metabolic rate will lead to storage of excess energy as fat, and obesity shall follow as the night the day. (I grossly oversimplify, you understand.) Thus, such studies require accurate, continuous metabolic data with excellent temporal resolution. Acute measurements of human metabolic rate are certainly informative and are briefly reviewed at the end of this chapter, but they are too burdensome and intrusive to be used for more than an hour or two at a time. The only method that delivers continuous metabolic data with high temporal resolution is flow-through respirometry, specifically using a room-sized chamber—a so-called room calorimeter (see Ravussin et al., 1986; Vioque et al., 2010 and references therein).

Room calorimetry is thus becoming increasingly popular. The number of room calorimeters around the world is rapidly expanding. An analysis of occurrence of the term "room calorimeter" in Google Scholar spanning the range 1988–2017 shows a steady, gently exponential increase in citations (Lighton, unpublished data) (Figure 12.1). What

Measuring Metabolic Rates: A Manual for Scientists. Second Edition. John R. B. Lighton, Oxford University Press (2019).
© John R. B. Lighton (2019). DOI: 10.1093/oso/9780198830399.001.0001

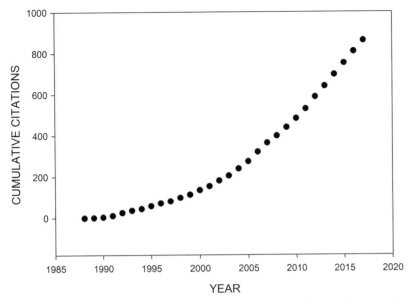

Figure 12.1 Number of publications in the Google Scholar database referring to "room calorimetry" from 1988 to 2017. A gentle exponential trend is evident.

follows is a brief and somewhat idiosyncratic synopsis of methodological issues related to room calorimetry, based partly on my own experiences in the field. There is a great deal of fear, uncertainty, and doubt concerning room calorimetry that in my opinion is unwarranted. I hope to show that room calorimetry is, in fact, simple to implement (just as it is also easy to complicate, as in most methodologies, but I am an evangelist for simplicity; to paraphrase Einstein, everything should be made as simple as possible, though not simpler; Einstein, 1933). General treatments of the topic can be found in Levine (2005), Sazanov and Schuckers (2009), Brychta et al. (2010), and, telegraphically, in Lighton and Halsey (2011) and Lighton (2012).

The idea behind room calorimetry is simple: Morph a mask to envelope the subject, add square corners, walls, a desk, a computer, a bed, an exercise machine, a latrine, and perhaps a view window; and voila, a respirometry chamber with all modern conveniences, also known as a room calorimeter. Most animals would go insane in such an environment, as you will observe at many zoological gardens. For most humans, especially those prone to development of obesity, it's become their natural habitat. As for the name given to the apparatus—well, it is not actually a calorimeter (Chapter 4), but historical inertia ensures the term's stability. "Indirect" calorimetry is implicitly assumed.

First, however, comes a fundamental decision: Should air be pushed into or pulled from the room?

The pull respirometry strategy, as employed with a conventional mask, makes a lot of sense in this application. Consequently, the majority of room calorimeters in use today are operated in pull mode, as Brobdingnagian masks. Of course, both push and pull mode systems are practical and are in use, and each has advantages and disadvantages as well as hybrid offspring; these are listed below.

Pull mode

A pull mode room calorimeter maintains a slight negative pressure with respect to its environment. This arises as a natural consequence of pulling air from a partially sealed room. The better the room's seal, the more negative the pressure. Where a leak occurs, the flow direction will be inward. This helps to ensure that negligible amounts of exhaled air will leave the room other than via its main exhaust port(s).

   Pull mode systems require a buffer zone of uninhabited air around the room calorimeter, or above it if the vertical walls are well sealed. The room draws its air from the buffer space, which must maintain as constant a gas concentration as possible for O_2, CO_2, and water vapor, unless the last species is removed rather than compensated for (see the section "Sample drying"). It follows that the buffer space should be as large as possible to minimize short-term fluctuations and should ideally be serviced by a heating, ventilation, and air conditioning (HVAC) system that draws air from outside the building. If the HVAC air intake is situated well away from pedestrian or vehicular traffic, so much the better. This makes the top floor of a multistorey building an optimal site for a pull mode room calorimeter. A low-resistance inlet is provided from the buffer space into the interior of the room calorimeter, typically through at least 2 m of 2- to 6-cm-diameter tubing to minimize the possibility of reflux from the interior to the exterior of the room in the event of large pressure pulses from other sources such as ill-behaved HVAC systems, the opening and closing of doors elsewhere in the building (if the buffer space is coupled to the building's interior), and strong gusts of wind.

   In practice, eliminating all leaks is virtually impossible. That is the bad news. The good news is that in a pull system, leaks in moderation are a *good thing*. The main inlet to the room from the buffer space, as outlined above, is simply a large, controlled leak. Leaks must, however, be minimized where they have significant area or occur in areas of the room calorimeter not in contact with the buffer space, especially if the air outside such areas differs significantly from the room's main source of incurrent air.

   In many ways, the arguments for maintaining a slight negative pressure within the pull chamber (or mask) are similar but inverted to those used in clean room technology, where contamination of a scrupulously clean area is avoided by maintaining a slight positive pressure. It surprises many people to learn that a rigorous standard against particulate contamination requires a differential pressure of only 5 Pa, or 0.5 mm of water column (ISO 14644-4). Most people are even more surprised to learn that this figure is, in fact, highly conservative. Experimentation at Lawrence Berkeley National Laboratory (Xu, 2007) has determined that rigorous isolation is, in fact, maintained at a differential pressure of only 0.025 Pa! This is a tiny pressure differential by anyone's standards; only 2.5 microns of water column, about the length of a typical bacterium.

The leakiness of a room's pull system can be monitored by measuring the differential pressure between the inside and outside of the room with a digital or fluid-based inclined manometer. The latter will give a simple, inexpensive, at-a-glance indication of a room calorimeter's pressure differential, but suffers from the disadvantage that its readings cannot be recorded together with the overall data stream from the room's respirometry system. Maintaining a continuous record of the differential pressure gradient between the outside and the interior of a room calorimeter is highly recommended.

The flow resistance of the main incurrent air source can easily be made adjustable via a butterfly valve or an iris mechanism, which allows fine control of the pressure differential at a given flow rate, assuming that the rest of the room is well sealed, which it should be.

Leaks in pull mode room calorimetry systems are advantageous in another respect. The inlet port to the room provides an ingress route for O_2 and an egress route for CO_2 in the event of a pump or power failure that will prolong the maximum acceptable "sealed" enclosure duration. This issue is addressed in greater detail in the section "Safety issues."

As shown above, provided that the leaks are small and only pull in air from a stable buffer zone, their effects will be negligible. But such is not always the case. Where blood sampling or other activities require breaching the room calorimeter's seal, such breaches are often from an adjacent occupied space that can differ in O_2, CO_2, and water vapor concentration from buffer-zone air. This may cause substantial discontinuities in the metabolic data. An air lock provides the obvious way of minimizing this effect for passing samples, literature, food, and beverages to and fro.

However, where human intervention with the room's occupant is required, as with taking blood samples, air locks are impractical. Flexible gloves, such as those used in glove-boxes, tend to be too thick to allow the sensitivity of touch required for blood sampling. In a pull system, unless double-sealed with a porthole cover, they may also tend to be inflated into the room because of the pressure differential and might unnerve sensitive subjects. Flexible iris seals behind a sealable porthole can reduce though not eliminate these problems (but see under "Push–pull mode" for an alternative approach).

The appropriate equations for calculating VO_2 and VCO_2 in a pull mode room calorimeter are 11.7 and 11.8, respectively. For conversion of VO_2 and VCO_2 to energy equivalents see equations 9.13–9.15.

Push mode

Considering the difficulty of guarding a pull mode room calorimeter's outer surfaces and ensuring that they are only bathed in stable and well-buffered air, a push mode room calorimeter appears an attractive alternative. Air can be pumped from a stable source such as outside the building, which may be at some distance from the room. Incurrent flow can be regulated at a rate affected, for purposes of calculation, only by the water vapor content of the incurrent air, somewhat simplifying the respirometry equations. However, push mode room calorimeters suffer from two main practical disadvantages.

The first disadvantage is safety. When a push mode room calorimeter suffers a power or pump failure, no significant air exchange occurs with its external environment. With a pull mode system, limited gas exchange may take place through the room's passive ingress port and other leaks, adding a small safety margin. Safety issues are addressed further below.

The main disadvantage of push mode room calorimetry systems is the possibility of inadequate recovery of the inhabitant's exhaled air. Considering that all practical room calorimeters will leak to some extent, a push system raises the risk that some portion of a subject's exhaled breath will leave the room by a route other than the main exhaust port that feeds the system's gas analyzers. In fact, I would go so far as to say that this unfortunate outcome is all but certain. The use of push mode room calorimetry should be entertained only by investigators with a Platonic conviction regarding the perfection of their room's construction and ongoing maintenance. The vast majority of push mode room calorimeters worldwide are to be found in Europe and Japan.

The appropriate equations for calculating VO_2 and VCO_2 in a push mode room calorimeter are 10.5 and 10.6, respectively. For conversion of VO_2 and VCO_2 to energy equivalents see equations 9.13–9.15.

Push–pull mode

In a so-called push–pull system, incurrent air is pushed into a pull mode room calorimeter at a rate sufficient to partially neutralize its negative internal pressure. Typically, the differential pressure between the room and its surroundings is measured, and a positive flow system is modulated via proportional–integral–derivative (PID) control to achieve a constant, small negative pressure within the room. The result is a massive increase in the overall complexity of the room's gas delivery and sampling system—to the extent that in one installation I am aware of, the marvelously intricate and substantial push–pull control systems are displayed in spot-lit splendor behind a bronzed glass panel for the admiration of visitors.

The major advantage of the push–pull approach is that maintaining a minimal negative pressure differential mitigates the effects of minor breaches of the room calorimeter's sealing system. For example, opening a sample port in the side of the room when taking a blood sample will not cause a sudden, major ingress of air through the port, carrying with it, probably, exhaled air from the person administering the bleed from an occupied space. In addition, there are safety considerations; if the pressure differential is minimized, then doors are easier to open. However, effectively identical results can be obtained with less fuss, complexity, and expense by simply reducing the inlet resistance of a conventional pull mode room calorimeter, and, if still deemed necessary, by installing pressure-relief mechanisms that can be readily operated in an emergency.

Advocates of push–pull room calorimetry (Moon et al., 1995) strongly recommend restricting air flow for a period immediately following enclosing the subject in the room, until the excurrent CO_2 concentration reaches 0.4–0.5 percent, and then maintaining it near that point by modulating the flow rate pulled from (and pushed into) the room. This adds substantially to the already considerable complexity of the room's control systems. The rationales for this approach, so far as I can determine, are (1) that CO_2 levels in inspired air have a physiological effect on subjects, from which it follows that to maintain comparability between investigations, the CO_2 level within the room should be controlled at similar levels; and (2) that the accuracy of CO_2 analyzers is rumored to be optimal near this concentration.

I find neither argument for what I will refer to as stenocapnic respirometry convincing. This is quite apart from the complexity, and thus potential for unexpected behavior or outright failure, inherent in such control systems. To address these points in turn, some of which are also relevant to conventional room calorimetry, let us first consider the argument for maintaining a constant CO_2 concentration.

The effects of CO_2

The purpose of deploying and using a room calorimeter is to measure the metabolic rate of a human being in their natural (*sensu lato*) surroundings. The lighting, temperature, and other physical and even aesthetic aspects of the room's interior environment are usually optimized to create a pleasant or at least neutral space. It seems obvious that O_2 and CO_2 concentrations should be likewise optimized to remain as close as feasible to normoxia and normocapnia.

This is especially important in the case of CO_2. As large terrestrial homoeothermic endotherms, humans receive most of their ventilatory cues from CO_2, not from O_2. An excess of CO_2 causes not only hyperventilation, but also pH and electrolyte alterations in body fluids. Neural function is soon compromised. Narcosis and other symptoms become obvious at a concentration of about 2 percent, and at high concentrations—several percent—death can result, even in healthy individuals. These dramatic effects have subtle precursors, however. Some authorities even hold that CO_2 concentrations below 0.2 percent can cause subtle affective and cognitive effects; the dysphoria, headaches, and mind-fog accompanying "sick building syndrome" have been linked to such CO_2 concentrations (Apte et al., 2000; Erdmann et al., 2002; Satish et al., 2012). Regulatory agencies have set the legal limit for long-term exposure to CO_2 at 0.5 percent for no more than 8 hours (ACGIH, 1991). Thus, for both scientific and regulatory reasons, the operators of room calorimeters should ensure that this concentration is never consistently exceeded. Maintaining CO_2 concentrations well below the 0.5 percent level may be required for pediatric, elderly, or diseased individuals. I believe that deliberately maintaining CO_2 levels close to 0.5 percent in the long term is both regulatorily and scientifically ill-advised, and, rather than increasing repeatability, may in fact reduce it because of individual variation in susceptibility to incipient hypercapnia.

Naturally, maintaining fixed and higher rates of flow will reduce the difference between incurrent and excurrent gas concentrations, and thus impose greater demands on the stability and noise specifications of the gas analysis system. Increasing the flow rate must necessarily make response correction trickier from the noise-filtration point of view (see the section "Response correction"), but, balancing this, it will reduce the magnitude of the response correction because of the room's faster time constant. Optimizing flow rate against analyzer noise and drift, taking the need for response correction into account, is an interesting subject that to my knowledge has not yet been rigorously addressed.

Next, we move to the purported benefits of stenocapnic respirometry in terms of analyzer accuracy. While the linearity of CO_2 analyzers should not be taken on faith (see Chapter 16), good-quality CO_2 analyzers are capable of delivering highly accurate results over a very wide range. More important, as outlined above, reaching such high excurrent CO_2 concentrations is undesirable for both scientific and regulatory reasons, and is in any event unnecessary if modern, well-linearized, and properly calibrated CO_2 analyzers are used. The typical room calorimetry flow rate of 30–60 L min^{-1} is probably inadequate to maintain CO_2 concentrations consistently below 0.5 percent with most subjects; flow rates of 80–100 L min^{-1} are more appropriate.

Response correction

Whether push or pull or other modes of flow-through respirometry are used, convective mixing within the room calorimeter must be extremely thorough. This is generally achieved either by using multiple fans or by using a fan coil unit which doubles as the heat exchanger for temperature control. (Incidentally, issues with "low recovery", i.e., underestimated EE measurements in room calorimeters, are almost invariably traceable to leaks in the HVAC or convective mixing systems.) The aim is to approach close to ideal mixing with no significant boundary layers, which requires a mixing rate of approximately half of the room's volume per minute. This is perhaps the most critical aspect to room calorimeter design. Ensure that no inadvertent leaks reduce the accuracy of the system; add a pump to draw air from, or pump air into, the room at a known rate; plumb the requisite gas analyzers; add a soupçon of signal processing, and the job is done. No air of mystery is required—or, shall we say, it is optional and more decorative than functional.

Why is thorough convective mixing an absolute requirement? If you are accustomed to dealing with animals in moderate-sized respirometry chambers with a time constant of a few minutes at most, you will be ill-prepared for the shock of seeing the raw data from your first room calorimetry run (always assuming that your system allows you access to the raw data). Far from rapidly reaching a plateau value, the O_2 and CO_2 traces will show a glacial change, lasting many hours, before even the semblance of a steady-state condition is reached. When the experimental subject runs on a treadmill, the O_2 trace will fall and the CO_2 trace will rise, in a gradual and seemingly linear fashion, until the running stops—quite unlike the clean, sharp plateaus seen with much smaller chambers.

The reason for this pathological behavior is very simple. If we assume a typical room volume and flow rate of 30,000 L and 80 L min^{-1}, respectively, the room's time constant is 30,000/80 or 375 min, which is over 6 h. Thus, 6 h after the start of a room calorimetry session, the excurrent gas concentrations will have reached only 63 percent (i.e., $1 - 1/e$) of their equilibrium levels!

Consequently, response correction (see Chapter 8) is an absolute necessity in room calorimetry. However, the challenge posed by the enormous time constant of room calorimeters is extreme. This is because the first derivative of the gas concentration traces must be multiplied by a huge amount—about 13,000 or more—prior to being added back to the original signal. Any noise in the derivative—and derivatives are notoriously noisy—will be likewise amplified and may overwhelm the metabolic signals. The noise can arise from the gas analyzers or from poor convective mixing—which is why excellent convective mixing is an absolute requirement in room calorimeters. Conventional smoothing techniques prior to derivatization, or smoothed derivative calculation algorithms (Savitzky and Golay, 1964), can reduce the differentiation noise to acceptable levels, but often at the cost of seriously damaging the temporal resolution of the system. If such techniques are used in the hope of resolving energy expenditure during a 20-min treadmill run, for example, the run may be temporally smeared into a smooth Gaussian curve without an identifiable plateau corresponding to energy expenditure during the run. To be of any use, response correction algorithms must be employed that allow a temporal resolution of the order of a minute or two—a seemingly impossible task, given a multi-hour time constant.

Solutions to this signal-processing problem are in a state of flux, but among the most widely used approaches is an ingenious algorithm refined by Nguyen et al. (2003), based on the pioneering work of Henning et al. (1996). In Nguyen et al.'s words, "Using the exact solution of the equations for steady state, the O_2 concentrations from the preceding 30 min period are fitted to two connected exponential segments, of variable length, using the least-squares method. The smoothed O_2 concentration and associated time derivative are then determined for the time point 15 min earlier and substituted into the respiration equations. The CO_2 concentrations are subjected to the same analysis. The process is repeated every minute, and the newly computed rates of O_2 consumption and CO_2 production, as well as metabolic rate, are then presented." This algorithm was obviously developed to extract energetic data during fairly brief ($\sim$ 20-min) exercise sessions with consistent energy expenditure, and works quite well for the purpose. Owing to its nature (iterative fittings of least-squares regressions), it does, however, create minor artifacts such as "divots" in short exercise plateaus, which Nguyen et al. (2003) refer to as an "M-shaped response," and which is clearly seen in Figure 2 of Brychta et al. (2009), together with a rather "steppy" response to small fluctuations.

Another approach, the central difference method (Sun et al., 1994), uses a discrete time derivative based on a 3-min window; in other words, and simplifying somewhat, the derivative for minute 2 is assumed equal to the mean value of minute 1 subtracted from the mean value of minute 3. This reduces noise and, relative to the Henning approach and its progeny, offers improved response to transient phenomena such as brief exercise bouts and makes fewer assumptions about the nature of the signal to be extracted, but has higher noise owing to its shorter window.

Another approach uses more sophisticated signal processing (Brychta et al., 2009), namely wavelet denoising, in conjunction with the central difference method in order to further reduce the derivative noise. Wavelet denoising is broadly related to Fourier transform analysis, and first decomposes the input signal into a series of frequency bands with characteristics determined by a "mother wavelet" (see references in Brychta et al., 2009). By applying variable thresholds to these bands derived from the nature of the noise to be contended with and then reassembling the signal, short-term noise is minimized yet response times are still quite rapid, which is vital for excellent system transient response. The method shows great promise, but is far from trivial to apply.

Finally, another approach that makes no assumptions whatever about the nature of the system noise and operates on a second by second basis was described by Melanson et al. (2010). I developed this "sliding polynomial" noise reduction technique for response-correcting data generated by Ed Melanson's 28,000-L room calorimeter within the constraints of a data analysis package (ExpeData) that I had designed for more generalized use. Essentially, the $(FiO_2 - FeO_2)$ and $(FeCO_2 - FiCO_2)$ vectors, which are acquired at 1 Hz, are fitted to high-order polynomials over a selectable window. The window is iteratively slid one sample futurewards, and a vector of the sums of the data estimates created by the central two-thirds of the polynomial fits is created. The summed vector is finally divided by the number of fits per sample (contained in another vector), yielding a vector of denoised data that follows short-term trends but has significantly reduced short-term noise. This vector is then derivatized and treated by the appropriate Z value derived from the time constant of the room (see Chapter 8). This method gives good results (Melanson et al., 2010), with good transient response, but it needs to be systematically optimized to further improve its noise performance. It is essentially a brute-force (and admittedly far less elegant) version of Savitzky–Golay smoothing and derivatization (Savitzky and Golay, 1964), which should yield equivalent results at much less computational expense. It would be particularly interesting to see this method compared against the wavelet-denoised derivation of the central difference method.

Sample drying

As in all respirometry, it is an absolute requirement that the dilution effect of water vapor must be eliminated from the measured concentrations of O_2 and CO_2 prior to calculating VO_2 and VCO_2. The protracted nature of room calorimetry runs makes chemical scrubbers less than ideal, so most room calorimetry analyzer chains use either thermal condensers or Nafion dryers to remove most of the water vapor from the sample (see Chapter 19). Neither method achieves complete desiccation, especially in the case of the widely used thermal condensers, which cannot reduce the partial pressure of water vapor below about 0.65 kPa, corresponding to a dewpoint of 1°C; see Figure 8.7, or even ~ 0.9 kPa (5°C). If the ambient WVP falls below this value, the resulting dilution of O_2, in particular, falls significantly. The result is an enrichment of O_2, a spurious elevated VO_2, and an equally spurious low RQ (see Melanson et al., 2010, for a discussion).

Nafion dryers (see Chapter 19) are a better alternative, but are complicated to implement because they require a countercurrent flow of gas at a very low WVP (such as pure N_2) or at a very low pressure.

A better solution, in my opinion, is to measure WVP and barometric pressure directly, and use Dalton's law of partial pressures to compensate for the dilution effect of water vapor mathematically, as outlined in Chapter 8 (especially the section on mathematical scrubbing), and also in Chapters 10 and 13. It is even easy to implement automated calibration of the WVP analyzer, as shown in Figure 10.4 and described by Melanson et al. (2010).

Analyzer drift

Response correction and analyzer drift are closely related issues. To put this in perspective, if the magnitude of response correction required for a typical small-animal system is scaled to a value of 1, for a room calorimeter it scales to about 2000. This, in turn, means that baselining the gas analyzers during the run is out of the question, because doing so would disrupt the excurrent gas traces and destroy the response correction. (As Chapter 8 explains, response correction algorithms are based directly or in practical effect on the first derivative of gas concentrations, so that any disruptions in those concentrations caused by anything other than the occupant of the room will lead to massive noise spikes that overwhelm the metabolic signal.) Thus, analyzer drift—primarily of O_2—and incurrent concentration changes—primarily of CO_2—cannot be accounted for.

Most room calorimetry systems use differential O_2 and CO_2 analyzers to partially address this problem. Some may quibble with the term "partially," but it is accurate. Even though differential analyzers should, in theory, only respond to differences between the reference (incurrent) and sample (excurrent) air streams, they are imperfect and will in practice drift. Periodic returns to identical incurrent air streams to track this drift are not possible in conventional room calorimetry systems because of response correction considerations. In order to minimize the contribution of analyzer drift to the analyzers' signals, it then becomes desirable to reduce flow rates to the minimum possible value that avoids the onset of hypercapnic symptoms in the subject. This, in turn, may necessitate making flow rate a variable parameter in order to maintain tolerable conditions within the room calorimeter during bouts of exercise. This was obviously part of the genesis of stenocapnic respirometry. This considerably complicates data analysis. Differential analyzers suffer from a further flaw: In the case of CO_2, where incurrent concentrations are variable and should be accounted for, data on incurrent concentrations (necessary for accurate VCO_2 calculation) are lost.

Schoffelen et al. (1997) describe one way out of this impasse. They interleaved measurements of incurrent and excurrent air with zero (N_2) and span calibrations, all of which are 1 min long. The resulting coverage of zero- and span-corrected gas concentrations from the incurrent and excurrent air streams was 48 min h^{-1}, or 80 percent. Accuracy, as determined by ethanol burns and CO_2 infusion, was very good. In practice, the temporal granularity of the system was 30 min, which, while adequate for integrative measurements, is too slow to acquire detailed data on the energy cost of short-duration activities.

I have developed a solution to the analyzer drift problem called "background baselining," which has been validated in a comprehensive trial (Melanson et al., 2010). The system accomplishes the seemingly impossible task of regularly baselining the gas analysis system without disrupting the gas exchange measurements (i.e., 100 percent coverage of the excurrent air stream) by utilizing two separate gas analysis chains (O_2, CO_2, WVP, and barometric pressure). Both gas analysis chains usually monitor the excurrent air from the room calorimeter. At intervals, however, one or the other gas analysis chain (never both) is directed to sample from incurrent air instead. The analyzers alternate these incurrent measurements under computer control. Care is taken to ensure that ample co-sampling (where both analyzer chains are sampling from excurrent air) takes place; see Figure 12.2 for an example of the principle. During analysis, the O_2 trace from each analysis chain is mathematically corrected for water vapor dilution (on the basis of the water vapor and barometric pressure data; see Chapter 8) and for variations in barometric pressure, then spanned to a fractional O_2 concentration of 0.2094, which is the worldwide O_2 fractional concentration standard in dry air and is more accurate than any readily available O_2 span gas (Tohjima et al., 2005). The excurrent sections of the two O_2 traces are then seamlessly combined to create a fully corrected but uninterrupted excurrent O_2 trace. An interpolated trace of the fractional concentration of incurrent CO_2 is then created (required for VCO_2 calculation in a pull system), and the delta O_2 and delta CO_2 traces are created from which the metabolic data are, in turn, calculated. The metabolic data from this system did not differ significantly from that measured in the same subjects at rest and during controlled activity using a highly accurate metabolic cart, and showed a temporal resolution, after response correction, of approximately 1 min during exercise bouts, in spite of the room calorimeter's Brobdingnagian time constant (Melanson et al., 2010).

Recent advances in networked data acquisition have allowed the deployment of "background baselining" room calorimetry systems to which external sensors can, in the future, be added and their data streams synchronized with the calorimetry data. Examples of projected sensors include activity, body mass, position, light level, food and water intake, and treadmill speed/inclination sensors, together with wearable sensors (heart rate, preprocessed accelerometry) that communicate via Bluetooth, Zigbee, WiFi, or other protocols. Because all data are acquired at 1 Hz from all sensors (with the accelerometry data processed and binned by the second for each axis), essentially perfect data synchronization is built into the system by default. Obviously, other consumer-level wearable sensors such as smartphones and smartwatches can also be used if they have sufficiently granular temporal resolution and are not locked into a walled garden from which time-stamped data cannot be exported.

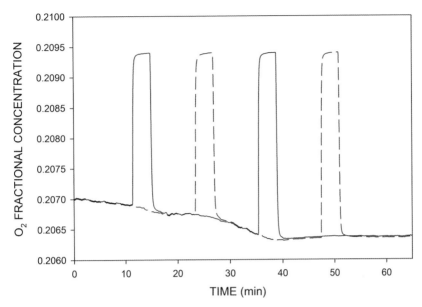

Figure 12.2 "Background baselining" in action. At the beginning of the recording, two O_2 analyzers are simultaneously monitoring the excurrent air from a room calorimeter. A little after 10 min into the recording, analyzer one (solid line) is switched to monitor incurrent (baseline) air instead. Analyzer two (dashed line) continues monitoring the excurrent air from the calorimeter. Analyzer one then returns to monitoring excurrent air; after several minutes, analyzer two is switched to monitor incurrent air, and the process repeats, with the analyzers alternating incurrent measurements, but with at least one, and usually both, analyzers monitoring the excurrent air from the calorimeter. Each incurrent air sample can be used to recalibrate the O_2 analyzers and essentially eliminate drift. The analogous process with CO_2 analyzer signals allows the creation of an $FiCO_2$ vector, which is essential for the accurate calculation of the VCO_2 in pull mode systems (eq. 11.8). After suitable corrections, the data vectors can be "stitched" together mathematically, greatly reducing typical sources of error in room calorimetry.

This "background baselining" technique is, of course, also applicable to multiple-animal metabolic measurement. Ideally, two gas analysis chains should be used for each animal, though with care this ideal can be relaxed somewhat to about $(N + N/4)$ analysis chains, where N is the number of the animals to be measured. If implemented for that purpose, it will finally allow metabolic phenotyping centers to obtain completely uninterrupted metabolic data from many animals simultaneously (see Chapter 13).

Validation of room calorimeters

Controlled burns of propane are usually carried out to validate the measurements made with the rooms. Ideally, the mass of the propane torch is continuously monitored and recorded, allowing simple conversion from the rate of mass loss to VO_2 and VCO_2. Propane burns usually show an O_2 and CO_2 recovery of 98 percent or higher and an RQ very close to the theoretically expected value of 0.60 (see Melanson et al., 2010 and references therein). Details on using propane and other burns for this purpose, as well as other approaches to validation, are described in Chapter 14.

Safety issues

Murphy's kingdom is any place where humans and machinery mingle. At some point, a room calorimeter's pump will fail or the power to its building will be cut. The question is not if, but when. In a worst-case scenario, a large person in a small room will fall asleep before the power fails and remain obliviously snoozing in a practically sealed room for several hours, perhaps overnight. The consequences could be unfortunate.

It is possible to mitigate this problem by designing fail-safes such as an opening in the wall of the room calorimeter—for example, a sampling port—that is normally held shut while power is available, but which opens if power or flow fails; a mechanism for unlatching the door if power or flow fails; or by alerting the room's occupant to leave via a battery-operated power or flow failure alarm. A CO_2 sensor that triggers an alarm condition if levels rise above a settable threshold is also a possibility.

Because any system that relies on human judgment is bound to exhibit sporadic irrationality, the most effective safety features are largely or completely automatic and multiply-redundant. For example, a battery-backed flow and power failure alarm coupled with an auto-dialer that calls several people in succession until a call is answered and a disable code is entered, and that is backed up by an automated door-unlatching or other seal-breaking system, is worth considering. Good-quality flow generators often have dry-relay contact breakers or other alarm interfaces.

Contacts intended for alarms should *open* when the alarm condition occurs. Such contacts are referred to as "normally open." For example, the dry relay contacts in a flow generator should close when the flow rate is within a specified distance of the setpoint. If the flow rate deviates from this condition for a prolonged period (typically a few minutes), or if the power fails, the relay will open and the attached alarm system will be triggered. Likewise, if a passing nitwit pulls out the cable to which the alarm is attached, an open circuit will result that will trigger an alarm. A system that denotes an alarm condition by closing a relay contact will not detect a power failure or other failure scenarios.

The flow generator's on–off switch should be difficult to activate accidentally; it may be worth considering using a "Molly guard" switch (named after a computer scientist's young daughter) surrounded by a bracket that must be deliberately opened in order to activate or deactivate the switch.

Any system for dealing with power or flow failure should of course be tested thoroughly, and on a regular basis.

Even in the absence of safety features, most room calorimeters are so large that it may take several hours for internal CO_2 concentrations to rise to dangerous levels. The danger is therefore not immediate, but that does not mean that it does not exist. Assuming

Table 12.1 The acute effects of CO_2 concentration on humans.

CO_2 concentration (percent)	Effect on humans
2–3	Shortness of breath, hyperventilation
5	Heavy breathing, tachycardia, sweating
7.5	Headaches, distorted vision, tachycardia, hypertension, dizziness
10	Nausea, vomiting, loss of consciousness

a generous CO_2 production rate of 1 L min^{-1}, a rather small room calorimeter with a volume of 20,000 L, and a CO_2 concentration of 0.5 percent prior to power failure, the occupant will show few obvious symptoms before $20,000 \times (0.020 - 0.005)$ L$^{-1} = 300$ min or 5 h. However, they may be seriously compromised if an overnight failure occurs, for example from 9 p.m. to 7 a.m., and the event is neither noticed nor mitigated by automatic safeguards. To put this in perspective, CO_2 concentrations above 4 percent are considered immediately hazardous to life and health (see Table 12.1).

Given that 4 percent is a typical end-tidal CO_2 concentration, it is difficult to imagine a higher CO_2 concentration developing in a room calorimeter, but anything approaching it will certainly produce profound ill-effects. A preventable accident could set back the field of room calorimetry by triggering legislative overreaction, and is easily avoided if mechanisms for coping with power and pump failure are planned and tested well in advance of the event.

SHORT-TERM (MASK) HUMAN METABOLIC MEASUREMENT

No chapter on human metabolic measurement using the room calorimeter is complete without a brief description of commonly used methodologies for acute measurement of human metabolic rates. Humans are large animals, and for many years their metabolic rates were primarily of interest to physicians and scientists concerned with human performance in the military or sporting arenas. Consequently, in the grand taxonomy of metabolic measurement, human metabolic measurement techniques branched off early and have maintained a separate evolutionary trajectory from mainstream metabolic measurement techniques in other fields.

As covered in earlier chapters, the universal practice in metabolic measurement using indirect calorimetry (except in the field of traditional human metabolic measurement) is to pull ambient air with known O_2 and CO_2 concentrations past the animal at a known flow rate. That flow rate is generally chosen so that the O_2 depletion generated by the animal is no more than about 0.5–1 percent. Thus, in essence, the breath of the animal being measured is greatly diluted by the air stream passing over it.

Because humans are large and tractable beasts, they can be trained to do things that most animals in their right minds would not consider doing, such as exhaling their breath into a sealed bag for later analysis. It is this charming tractability of humans that has set human metabolic measurement onto its own evolutionary trajectory, separate from the methodologies that must be utilized when dealing with creatures (even) less amenable to reason and persuasion.

Most of the measurement techniques used for humans are derivatives of the classic work pioneered by Haldane, Douglas, Barcroft, Zuntz, and others in the late nineteenth and early twentieth centuries. This is not the place to describe the development of the field in detail, nor to describe the contributions made by each of these pioneering scientists (though it is worth drawing your attention to Gunga (2008) for an excellent biography of Nathan Zuntz). Rather, I am concentrating on the broad principles of measurement that they developed.

In essence, the volume of air breathed out by a human volunteer is measured over time, either directly or by using a device such as a spirometer. The O_2 and CO_2 concentrations of this expired breath are measured, and the metabolic rate of the subject is calculated from the O_2 depletion and CO_2 enrichment in the undiluted expired air, and the volume of that air produced per unit time. The expired air can be either collected directly, for example in a Douglas bag, or sampled using a line leading to the tubing that houses the spirometer. In either case, it is important to realize that the sample being analyzed is either bulk human expired breath (in a Douglas bag or obtained using a one-way valve in conjunction with a spirometer) or a breath by breath fluctuation between expired and inspired gas concentrations.

Because human end-tidal O_2 and CO_2 concentrations are so far removed from ambient values, this approach makes minimal demands on gas analyzers, other than for speed if the O_2 and CO_2 concentrations of each breath must be tracked as in the second technique described above. This is in marked contrast to the analyzer characteristics required for more conventional flow-through respirometry, which accentuate low noise, low drift, and high resolution, but do not require very high speed because the time constant of the respirometry chamber renders high speed largely irrelevant (though it can easily be implemented mathematically, as described in Chapter 8, provided good-quality analyzers are used).

The Douglas bag technique is an excellent didactic tool but is no longer relevant to most research situations. The practice of measuring breath volume using a spirometer and sampling air from alternating inspired and expired air, or of sampling only expired air via a one-way valve, is still a very common methodology in human metabolic rate measurement. Variations of this general technique are usually referred to as "breath by breath" measurement. Thanks to aggressive marketing by companies making devices that utilize this technique, many physicians and scientists assume that "breath by breath" measurement is the last word in accuracy. Such claims should be taken with a mole or two of salt, for the following reasons.

First, "breath by breath" analysis depends on accurate volumes measured by a spirometer or other flow or volume measurement device. In fact, these volumetric devices are greatly affected by air temperature, water vapor content, air viscosity, and barometric pressure, not to mention the highly dynamic and nonlinear flow profiles found in tubes attached to masks. Thus, their error budgets can be expected to be greater than that of a typical mass flow meter which directly measures the molar, STP-corrected flow flowing down a constant pressure gradient generated by a pump. Because metabolic measurement utilizing gas concentration changes is critically dependent on flow accuracy, "breath by breath" analysis is intrinsically less accurate than measurement techniques utilizing mass flow meters, some of which are capable of much better than 1 percent of reading accuracy (see Chapter 17).

Second, because there is something of an arms race (also known as the Red Queen race) between vendors of different human metabolic measurement systems (often referred to as metabolic carts or simply "met carts"), the accuracy specifications cited by the manufacturers are often, to be charitable, wildly optimistic. This applies to the flow measurement parameters mentioned above, and also to the gas analyzers, where claimed accuracy levels of 0.03 percent for CO_2 analyzers are not uncommon. In fact, as discussed in Chapter 16, such accuracy levels are simply impossible and are derived by creatively misinterpreting the meaning of the word "accuracy." For example, an accuracy of 0.03 percent is claimed by defining a measurement as being within 0.03 percent of 1 percent when measuring a 1 percent span gas, which, first, is actually an accuracy of 3 percent, and second, incorrectly assumes that the span gas is perfectly accurate.

Third, as shown in Chapters 9–11, accurate O_2 and CO_2 concentrations must be exactly time-matched when calculating VO_2 and VCO_2 because CO_2 enrichment dilutes O_2 and O_2 consumption enriches CO_2. When measured by a serially connected gas analysis chain, these concentrations cannot be assumed to be synchronous, especially after passing through a desiccant system, and because the kinetics of O_2 and CO_2 analyzers are often quite different. Attempting to apply cross-corrections to rapidly varying, out-of-phase dynamic signals measured by two analyzers in series and with different response characteristics is an extremely difficult task, which as far as I can ascertain (because no raw data are available from any such systems) is not even attempted. In fact, most "breath by breath" systems actually use mixing chambers to stabilize gas concentrations prior to measurement, thus undermining their "breath by breath" claims if you look under the hood.

Fourth, the requirement to measure expired breath volumes imposes a concomitant requirement to achieve an extremely tight fit between the subject's mouth and the spirometer. This can be uncomfortable and invasive, especially because most systems require that the nostrils of the subject are pinched shut by an even more uncomfortable and invasive nose-clip.

Thus, while I admire the marketing bravura behind it, I regard "breath by breath" metabolic analysis as a less accurate and more cumbersome variant of conventional flow-through respirometry. If conventional flow-through respirometry, or as some call it "dilution" measurement, is used instead, flow measurement can be made simpler and more accurate, and the diluted expired air is far easier to analyze in the absence of rapid dynamic variations and phase shifts. This assumes, of course, that gas analyzers with high resolution and low noise are available because the absolute magnitude of the gas signals will be approximately ten- to twenty-plus-fold smaller. "Met cart" gas analyzers typically have a resolution of 0.01–0.1 percent at best, adequate for measuring directly expired air with acceptable accuracy. However, with the dilution technique, resolutions of the order of 0.001 percent are required at sensible flow rates, which is a different ball-game. Fortunately, that degree of resolution is *de rigueur* in research-grade flow-through respirometry, making equipment designed for such applications ideal for measuring human metabolic rates using the dilution technique.

Some "met cart" manufacturers have switched to using their own variant of the dilution technique, but considering the low resolution of their gas analyzers they are forced to use low flow rates, typically 20 L min^{-1} for resting metabolic rates. While this yields a large O_2 and CO_2 differential that is easily measurable, it also creates an uncomfortable and stale environment for the human subject. Subjects with very high resting metabolic rates require an elevated flow rate, while those with unusually low metabolic rates may require adjustment to even lower rates.

My own testing using research-grade gas analyzers has shown that excellent results for resting human metabolic rates can be obtained with flow rates in the range of 80–200 L min^{-1}, which for human subjects wearing a mask, hood, or canopy feels very much like typical ambient conditions. There is no need to adjust the flow rate to accommodate normal resting metabolic rates. Essentially, the equipment required for human metabolic rate measurement is identical to that described above for room calorimetry, though "background baselining" is overkill for this application. If the system is suitably configured, reliable O_2 and CO_2 incurrent concentrations can be obtained automatically in ~ 15 sec per reading (see Chapter 13 and especially Figure 13.1) and can be an essentially invisible aspect of system operation even when using a single analyzer, relieving the operator of the requirement to make explicit baseline measurements. Such a system can also be rendered independent of water vapor scrubbers by utilizing mathematical correction for water vapor dilution (see Chapter 8). If equipped with a flow generator covering the range of approximately 50–500 L min^{-1}, the system can reliably measure subjects ranging from sedentary children to elite athletes on a treadmill or bicycle ergometer.

The overall result of applying modern metabolic measurement techniques to human metabolic measurement is an increase in accuracy, a reduction in subject discomfort, greater ease of use, and (if desired) greater flexibility in the retention of raw data and improved traceability of the final results.

13

Metabolic Phenotyping

BACKGROUND

The explosive growth of biomedical research over the past half-century has led to a bottleneck in obtaining data from large numbers of experimental animals, particularly with respect to metabolic parameters, activity levels, and food and water intake amounts and patterns. Making such measurements on many animals quickly and efficiently required techniques and approaches quite different from those prevalent in the comparative physiology community, which had been quietly making demanding metabolic measurements for many years on small numbers of diverse animals. Therefore, large-scale biomedical metabolic phenotyping systems developed along a separate evolutionary arc, and in a distinctly different conceptual space, from the norms and expertise painfully acquired by comparative physiologists.

There are several consequences that flow from the fact that the first firms to develop biomedical metabolic phenotyping systems used an engineering rather than a scientific or biologically aware mind-set. This is not necessarily a bad thing, except that the biology of the animal being studied was often considered last, if at all, among the design priorities of such systems. This has led to several trickle-down effects—among them, use of low flow rates to maximize the signal from gas analyzers, minimizing chamber volume to attain reasonable time constants at low flow rates, and the creation of stressful, unnatural environments devoid of bedding or enrichment for the animals, which is now known to cause corticosteroid induction with its panoply of metabolic and other side-effects. These first best attempts made incorrect, nonbiological assumptions about the relation between metabolic parameters and body masses, leading to attempted normalization of energy expenditure via simple division by body mass or other equally ill-advised schemes (see Chapter 15). The effect of thermal parameters on the metabolic rates of small model animals was unfortunately not well understood even within the biomedical research community, so there was little attention paid to—for example—the ambient temperature at which measurements were made, and thermal metabolic effects may have been exacerbated by the use of convective ventilation to the cage system to improve gas mixing.

It is unfortunate that communication between the biomedical and comparative physiological fields has historically been absent at worst to poor at best, or some of these

Measuring Metabolic Rates: A Manual for Scientists. Second Edition. John R. B. Lighton, Oxford University Press (2019).
© John R. B. Lighton (2019). DOI: 10.1093/oso/9780198830399.001.0001

problems might have been addressed sooner. A few researchers such as Gerhard Heldmaier and John Speakman and other pioneers tried to bridge the divide but faced long odds. It is a sad fact that some of the data in the scientific literature derived from metabolic phenotyping systems may prove difficult and, in some cases, impossible to interpret correctly as a result of the problems outlined above.

At the time I was drafting the first edition of this book in 2006, the situation seemed so hopeless that as a comparative physiologist I elected to shy away from discussing bio-medical applications of metabolic measurement, limiting myself to a few comments on mass scaling and temperature effects in a chapter on data presentation and analysis. These comments on analysis and interpretation are now further expanded in Chapter 15, and a slew of recent treatments in the biomedical literature suggests that a set of best practices is being established—Kaiyala et al. (2010), Butler and Kozak (2010), Choi et al. (2011), Tschöp et al. (2012), Speakman (2013), Meyer et al. (2015), and Lighton (2017).

Consequently, in this chapter I will concentrate primarily on methodological rather than interpretational matters—see Chapter 15 for those. I have spent many years deeply involved with the analysis and design of metabolic phenotyping systems, both from the point of view of a practicing scientist and metabolic measurement specialist, and from the practical perspective of designing systems to address my concerns and those of my scientific peers in the biomedical community while walking the tightrope between ease-of-use and analytical flexibility. It is impossible for me to write this chapter without being accused of bias, for which I make no apology—unless of course I can be shown to be mistaken in any of the points raised below. Feel free to vent on Twitter @sablesys should you find any such mistakes!

This chapter should be regarded as complementary to the sections in Chapters 10 and 11 on measurements from multiple animals ("Extending to multiple chambers"), which deal in more basic detail with plumbing and so forth; to Chapter 8, which deals with the temporal limitations imposed on metabolic measurements; and to Chapter 9, which deals with the translation of indirect calorimetry data in energy equivalents. I will make frequent reference to subjects covered in detail in other chapters as well.

METABOLIC MEASUREMENT GOALS AND METHODS

The ideal multiple-animal metabolic measurement system provides continuous, immediate, accurate measurements from many animals simultaneously without any interruptions. This is the ideal and, as with most ideals, it is honored more in the breach than the observance. Rather, animals are usually sampled sequentially using a multiplexing approach, interleaved with measurements of incurrent air. Some obvious questions are: Does the multiplexed approach yield data acceptably equivalent to continuous sampling? And, when is it advantageous to sample continuously from each cage, using a separate analyzer chain for each cage? Does continuous sampling offer a temporal resolution advantage? Surprising answers to these questions are explored in the section "Continuous sampling."

Multiplexed sampling

In practice, the most common approach to metabolic phenotyping multiplexes (sequentially switches) the gas streams from multiple cages through a single gas analyzer chain. This approach is outlined in Chapters 10 and 11 ("Extending to multiple chambers"). The excurrent air from each cage is briefly sampled, creating a snapshot of gas concentrations from which VO_2 and VCO_2 can be calculated. Figure 13.1 shows a simple example.

Let us define some important terms. The number of cages in a metabolic phenotyping system is denoted by N. The time required to obtain an accurate reading of gas concentrations after switching the gas analyzer chain to a new sample (cage or baseline) is the *dwell time*, D. The number of cages measured per baseline measurement is called the *interleaving ratio*, I (for example, if a baseline is measured every four cages, the interleaving ratio is four). The parameter of greatest interest to most researchers, of course,

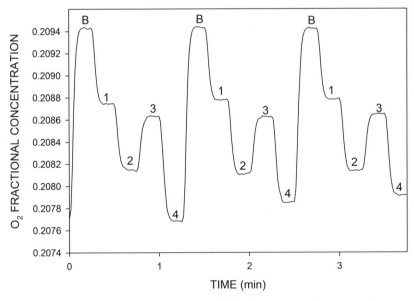

Figure 13.1 Demonstrating fast multiplexing of a fuel cell oxygen analyzer in a pull mode metabolic phenotyping system (Sable Systems Promethion) between four cages, showing Z-corrected data from a C57BL/6J mouse. The system uses a dwell time of 15 sec per cage (numbers 1–4) and interleaves a baseline measurement (B) between each set of four cages, giving a total cycle time of $15(4 + 1) = 75$ sec. The occupant of cage 1 has the lowest, and that of cage 4 the highest, metabolic rate. Similar data are available in the system for CO_2 emission and water vapor. Data are from a metabolic phenotyping course using four out of a possible eight cages for the system.

is the duration between successive measurements of any given cage in the system; this is the *cycle time*, C. Finally, in some systems, sampling duties are shared between multiple gas analyzer chains. The *analyzer number* (A) is usually one, but can be two or more.

It is immediately obvious that multiplexing methodology involves significant tradeoffs. First, best practice dictates that the more baseline (incurrent) samples that are measured between the cages (the lower the interleaving ratio), the more accurate the results will be. However, note that this increase in accuracy will be offset by an increase in cycle time. Second, the more cages are multiplexed, the longer the duration between measurements on any given cage (the cycle time). Third, the more rapidly the cages are multiplexed (the lower the dwell time), the faster the cycle time; but this is offset by reduced accuracy if the gas analyzers do not have time to equilibrate to the concentrations they are measuring.

To consider multiplexed metabolic phenotyping more quantitatively, let us generalize the problem in terms of the first two tradeoffs: first, baseline-derived accuracy versus an increase in cycle time; and second, increased N (cage number) versus cycle time.

Baseline measurement

For the greatest accuracy in gas exchange measurement, baselines—in other words, samples of incurrent gas concentrations—must be measured frequently. In contrast, for the maximum temporal fidelity of metabolic measurement, they should be measured as infrequently as possible. Taking the simplest case, where "spot" measurements are made in sequence for the shortest possible dwell time, we can interleave successive excurrent air streams from successive chambers (of total number C) with baselines. Typical interleaving ratios will range from 1 (each chamber reading is accompanied by a baseline reading) to N (baseline readings are taken only after a full cycle through all N chambers). Of course, larger interleaving ratios are possible, but these require irrational faith in the stability of incurrent gas concentrations (particularly CO_2) and of the analyzers themselves.

For this scenario,

$$C = \left(N + N/I\right) \times D \qquad (13.1)$$

For a given system, accuracy is inversely proportional to the interleaving ratio. Meanwhile, the throughput of the system is maximized if baselines are not measured other, perhaps, than at the start of measurements. Yet this is clearly inadvisable (though with rare exceptions this is generally the case with human room calorimetry systems; see Chapter 12). Discounting that extreme case, throughput is directly proportional to the inverse of the above ratio, ranging from N at I = 1 down to 1 at I = N. Fortunately there are strategies for ameliorating these tradeoffs, as you will see later.

For a practical example, consider a metabolic phenotyping system that measures 16 animals. If we assume a fast but conservative 30-sec dwell time per multiplexed gas sample, a complete cycle of measurements would take 16 min at an interleaving ratio of 1, and a much speedier 8.5 min at an interleaving ratio of 16. If we know that the combination of incurrent gas concentration changes and analyzer drift will not cause an unacceptable error over a cycle time of 8.5 min, then this baselining ratio is reasonable. Alternatively, adding just one extra baseline (interleaving ratio 8) will increase the cycle time by less than 10 percent to 9 min, while halving the interval between baselines from 8 to 4 min doubles the precision with which baseline corrections can be made in the presence of drift. This throughput could be doubled, or the cycle time halved, by reducing dwell time to the edge of feasibility (about 10–15 sec; see Figure 13.1), but this may introduce errors of its own unless implemented with care. Note that the dwell times I refer to are for state-of-the-art systems that are far faster than most on the market. Dwell times of 2–3 min or more are common among older systems that are still in wide use around the world, leading to cycle times of as much as an hour for systems with large numbers of cages. Such glacial cycle times are justifiable if the sole purpose of the system is to estimate overall mean 24-h energy expenditure, but they are of only marginal utility for distinguishing between nocturnal and diurnal energy expenditures. For capturing data on transient metabolic events, such as those accompanying bouts of activity, they are entirely useless.

Increasing the number of analyzers (A) is a partial solution for decreasing cycle times in large systems. If multiple analyzers are used, the above equation becomes:

$$C = \left(\left(N + N/I\right) \times D\right)/A \qquad (13.2)$$

There is another, less obvious advantage to the multiple-analyzer approach. If a system is expanded, for example from 8 to 16 or 24 cages, and if each analyzer chain is configured to multiplex samples from (for example) eight cages, then the cycle time of the system is not affected by the increased cage number. This yields greater consistency of temporal resolution across experiments with different sample sizes. It is, however, important that the analyzer chains be scrupulously matched. If multiple analyzers are used, control runs with a uniform population of animals should occasionally be checked for any by-analyzer effect. As an added precaution, different experimental groups should be randomized across analyzers.

Multiplexing speed

Next we consider the speed of multiplexing (which is inversely related to dwell time). It is obvious that without further data processing, the rate-limiting steps are the response speed of the gas analyzers, plus the time taken to convey a gas sample to the gas analyzers, to which must be added delay caused by the washout characteristics of the water vapor scrubbing system, if any. In the case of fuel cell O_2 analyzers, it may take a minute or so for full equilibration of the analyzer to the new sample (but see Chapter 8 for methods of increasing equilibration speed by applying the instantaneous transformation); this can be improved in terms of native analyzer response speed by using paramagnetic or zirconia-cell O_2 analyzers which are inherently fast (though they have their own issues and tradeoffs; see Chapter 16), in which case the scrubber becomes the main limiting factor.

The time taken to convey gas samples to the gas analyzers can be reduced to near zero by intelligent system design. In classical push mode systems, the gas sample to be analyzed is pulled from the cage at the start of the sampling period. Because the cage can be up to several meters away from the gas analyzing apparatus, it may take 2–3 min before a stable gas concentration arrives at the gas analyzers. Most push mode metabolic phenotyping system manufacturers do not include this delay in their technical specifications. In contrast, some modern systems use pull mode respirometry, in which gas samples are pulled at full speed from all cages simultaneously to the immediate vicinity of the gas analyzers, resulting in a negligible delay when switching the analyzers from one cage sample to another.

Water vapor scrubbers, generally—if mistakenly; see Chapter 8—regarded as necessary to eliminate water vapor dilution of respiratory gases, are generally based on chemicals such as Drierite, on selective membranes such as Nafion, or on thermal condensation (see Chapter 19). Chemically based scrubbers tend to be rather large in volume because of the long duration of typical metabolic phenotyping experiments and can significantly slow down response times. They may also introduce unacceptable phase lags between O_2 and CO_2 concentrations. Thermal condensers have a lower interior volume, but they do not dry the air below a water vapor pressure of about 900 Pa in typical metabolic phenotyping systems. Thus, they can cause major errors if the dewpoint of the ambient air drops below *c.* 5 °C, which corresponds to that water vapor pressure (see Chapters 8 and 12; also Melanson et al., 2010). Water vapor pressures in that region are not especially likely, as they correspond to relative humidities of ~ 35 percent or below at typical room temperatures, but may occur during certain experiments (see Figure 8.7 for a real-world example). Thermal scrubbers accumulate a reservoir of condensed water, which some assume will evaporate into the air stream if incurrent water vapor pressures drop below the device's condensation temperature. Obviously, however, this is only a short-term solution. The presence of liquid water is problematic in any event because of possible interactions between the water and CO_2 in which the condensed liquid may act as a CO_2 capacitor and cause dynamic RQ distortion. In addition, calibration gases are always dry and thus will give misleading results unless they are pre-equilibrated to the chiller's dewpoint. The theory is that condensation already present in the chiller will hydrate the calibration gas to the same water vapor pressure as sample air, thus paradoxically using a desiccator as a humidifier. Nafion dryers (see Chapter 19) tend to be mechanically complex and rely on the low water vapor pressure, or low absolute pressure, of a dry or low-pressure carrier gas into which water vapor can diffuse. Often this reliance is justified; but sometimes not, depending on the mechanical health of the system, which can be difficult to ascertain with confidence.

By using Dalton's law of partial pressures, it is possible to bypass these problematic technologies with their volume and response time penalties, edge-case breakdowns, and mechanical reliability issues. As discussed and validated in Chapter 8, it is merely necessary to measure water vapor partial pressure (see Chapter 16 for measurement principles) and barometric pressure, then use simple algebra to make the necessary correction (eq. 8.7). See Chapter 8 for a comprehensive validation of this technique using metabolic phenotyping system data. This technique assumes that the O_2 and CO_2 analyzers in the system are not affected by the presence of water vapor, other than via its dilution effect. Economy CO_2 analyzers that utilize a single CO_2 absorption wavelength may be susceptible to crossover infrared absorbance from water vapor. In the case of paramagnetic O_2 analyzers, water vapor may interfere with O_2 readings because it is diamagnetic, and thus opposes the paramagnetic effect of O_2. In the case of zirconia O_2 analyzers, water vapor is generally considered deleterious to the sputtered metal electrodes responsible for communicating the Nernst potential of the red-hot zirconium oxide cell to the outside world. Fuel-cell O_2 analyzers, in contrast, exhibit no measurable interference from water vapor other than via its dilution effect.

It is worth mentioning that for O_2 analyzer calibration, utilizing ambient air in an animal facility after appropriate correction for water vapor pressure dilution and barometric pressure yields results anywhere from 7 to 1000 times as accurate as even the most accurate commercially available bottled O_2 span gases (see Chapter 16 for a detailed discussion of this point).

Water vapor scrubbing strategies aside, multiplexing speed can be greatly increased by intelligent system design. First, applying response correction to the gas analyzer signals (see Chapter 8) results in a dramatic reduction in equilibration time with a consequent increase in switching frequency and reduction in cage-cycling time, *provided* the analyzers have a low intrinsic noise figure. Figure 8.4 furnishes a dramatic example of response correction, in which the response speed of a fuel cell analyzer becomes indistinguishable from that of a heated-zirconia analyzer, generally regarded as the fastest available (but which can be damaged by water vapor and which overestimates VO_2 in the presence of volatile organic compounds because it combusts them, making it a dubious choice for metabolic measurement). Figure 13.1 illustrates the point with an actual, rapidly multiplexed O_2 signal from a metabolic phenotyping system, acquired at 1 Hz. Again, this technique depends on low levels of analyzer noise because a substantial multiple of the derivative of the analyzer's signal is added back to the original signal; and the derivative is exquisitely sensitive to noise (Lighton and Halsey, 2011; Lighton, 2012).

Temporal resolution

To the naïve scientist, it is self-evident that to obtain the best possible temporal resolution of an animal's rapidly fluctuating metabolic signal, a very rapid cycle time or—preferably—a continuous metabolic measurement system is required. In practice, the situation is more complex and subtle. As outlined in Chapter 8 and further discussed below, the temporal resolution of a metabolic measurement system is limited by the time constant of the chamber containing the animal, which in turn is a function of the volume of the chamber and the flow rate passing through it. Given a typical chamber volume of approximately 8 L STP and a typical flow rate of ~ 0.4 L min^{-1} in most metabolic phenotyping systems, we see that typical time constants are > 20 min. Yet, as we learned in Chapter 8, it is impossible for the metabolic signal to change by more than ~ 63 percent during this time interval. Thus, sampling the air emerging from the cage at a time interval significantly faster than the cage time constant yields rapidly diminishing benefits. We will analyze the implications of this insight in more detail in the section "Continuous sampling."

The goal when multiplexing chambers should be to attain a cycle time comparable to, and preferably substantially less than, the time constants of the individual cages. In an ideal world, multiplexing should be fast enough that the concentration changes in gases pulled from the cages is minimal across a single cycle time. Again, in the case of a cycle time equal to the chamber time constant, the concentration change can be up to 63 percent (see Chapter 8). I suggest from empirical experience that cycle times should, where possible, not much exceed 50 percent of the time constant of the chambers in the system (see also Lighton, 2015). This is in accordance with the Shannon–Weaver model of communication (Shannon and Weaver, 1963), which dictates that a waveform can be adequately captured if it is sampled at twice its characteristic frequency—or, in the context of our discussion, at half the duration of the chamber time constant; and, as a corollary, that sampling more frequently brings little if any benefit. Thus, in a paradoxical fashion the temporal resolution limitation imposed by the cage time constant becomes our friend when viewed in the context of multiplexing, in that it imposes a limit to the benefits accruing from faster metabolic sampling. This may allow cautious use of response correction (see Chapter 8) even in metabolic phenotyping systems that are multiplexed, provided that a suitably sophisticated interpolation algorithm is used between individual data points. This point is further discussed below.

Extraction of metabolic data from multiplexed systems

Because the concentrations of the various gas species are interdependent, and these concentrations can fluctuate rapidly during multiplexing operations, calculations based on these concentrations must be implemented with care. Separate response corrections are required for O_2, CO_2, and WVP analyzers, and the signals from each category of analyzer must be temporally synchronized using a sliding correlation algorithm prior to further signal processing (see Figure 10.4). The object of the exercise is to extract a stable point from each switched concentration for further calculation. Various possibilities for extracting such points exist; in my opinion, the most robust is to choose a contiguous subset of concentrations with the lowest variance and extract the mean of each subset. The individual extracted data points can then be used in calculations as detailed in Chapters 10 and 11, which will not be further described here.

The result of these calculations will be vectors of VO_2, VCO_2, and RQ (together, in a system measuring water vapor pressure, with whole-animal VH_2O; see Chapter 8 for a discussion and validation of mathematical water vapor dilution correction) separated in time by the cycle time of the multiplexed system. These can be presented in tabular, graphic, or both forms as required.

Synchronizing with nonmultiplexed sensors

In practice, most metabolic phenotyping systems also acquire data from other sensors, for example the X, Y, and Z position of the animal, the number of beam-breaks caused by the animal in each axis (see Chapter 18), running wheel revolutions, and food and water intake. In many systems, data from these subsidiary sensors are acquired using stand-alone programs which may or may not produce outputs that can easily be synchronized with the metabolic data; for example, they may be binned quite differently.

In systems that record raw data at a high rate (see the section "Traceability"), it is possible to continuously record data from all sensors in a large multicage system together with the raw, multiplexed gas exchange data. This provides all the data required to ensure essentially perfect synchronization. However, as alluded to above, valid metabolic data points for a given cage may be separated in time by up to several minutes (the cycle time of the multiplexed system). How are the multiplexed data, which are discontinuous and at a much lower bandwidth with regard to each cage, to be synchronized with the continuously acquired, high-bandwidth data from each cage's other sensors?

One solution to this quandary is to create multiple VO_2 and VCO_2 channels, one for each cage, and then populate them by interpolating between the valid metabolic data points for each cage. This is my best recommendation for multiplexed systems and creates a pseudo-continuous record that is synchronized with the other sensors. These vectors can then, if deemed necessary, be left-shifted by the time interval empirically determined to elapse between activity in the cage and detection of the gas exchange signature of such activity, ensuring practically perfect synchronization. In the case of high flow rate, pull mode systems, this time lag is usually negligible, and correction is seldom required.

Of course, the seemingly continuous metabolic data for each cage is a cruel illusion but may suffice. Figure 13.2 shows an example in which a pseudo-continuous signal is created from multiplexed data. Note that the relationships between the interpolated line and the raw gas analyzer readings from which it was derived are clearly visible, allowing the quality of the data to be directly evaluated. Such evaluation is not possible if only spot readings, rather than a detailed time-course, are available for inspection and auditing. For this reason, I am a strong advocate for recording all data, including the raw gas analyzer data, at the maximum sampling rate used by all of the other sensors in the system, for example 1 Hz. Certainly, this produces large data files, but we should now regard storage capacity as infinite and govern ourselves accordingly. The added flexibility and versatility of data analysis, unconstrained by Procrustean a priori decisions that may prove to be wrong in retrospect, more than makes up for the trivial added storage space required. Spot readings without including the raw time-series data from which they were derived do not constitute raw data in any meaningful sense of the term.

If interpolation is used, the interpolation algorithm should be chosen with care. Arguably the best such algorithm was originally designed for computerized graphics generation and is known as the Catmull–Rom spline (Catmull and Rom, 1974; parenthetically, Ed Catmull is currently the president of the Pixar and Walt Disney animation studios). It is unusual because it passes exactly through each sample point and utilizes tangent vectors to approximate a "best guess" at the underlying nonlinear continuity of the data. The interpolated line in Figure 13.2 was created using a Catmull–Rom spline.

One weakness of the interpolation approach is that response correction (Chapter 8) cannot always be meaningfully applied to the resulting data, so the metabolic record is temporally compromised by the cage's time constant. This time constant can be quite considerable. Most metabolic phenotyping systems use very low flow rates because they use gas analyzers of indifferent quality and the manufacturers (and users) are tolerant of the resulting penalty in temporal resolution. Even when high-quality analyzers are used, a typical cage volume of 8 L STP from which air is pulled at 2 L min^{-1} STP yields a time constant of 4 min. It can be argued that in a multiplexed system, this approximates the cycle time and that greater temporal resolution may be redundant. Nevertheless, where possible, I advocate minimizing cycle times relative to chamber time constants.

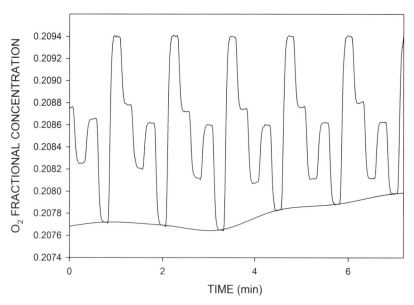

Figure 13.2 Creating a pseudo-continuous gas concentration record from multiplexed data. Data are as in Figure 13.1, but over a somewhat longer time interval. The line at the bottom of the recording shows the interpolated O_2 concentration data for cage 4. A similar interpolation works for VO_2, VCO_2, RQ, and so on. The pseudo-continuous line is now at the effective sampling rate of the raw gas concentration data and the cage sensors, which in this case was 1 Hz. Data are from a Sable Systems metabolic phenotyping course.

Another, inverse solution is to bin the sensor data to match the gas exchange data for each cage. Provided that all raw data are preserved, either or both approaches can be used.

Continuous sampling

Chapters 10 and 11 describe the basic methodology for implementing continuous respirometry using push or pull flow generation, respectively. The term "continuous" is of course incorrect because periodic measurements of baselines (incurrent gas concentrations) are essential. During these periods no data are acquired from the animal being monitored. In most cases this is not a major drawback, because baselining will typically account for no more than 5 percent of the monitoring time-budget. Because the duration of each baseline can be substantially less than the cage time constant, causing the probable deflection of the primary signal during the baselining event to be minor to negligible, it is possible by using suitable software to interpolate across the baselines, yielding a pseudo-continuous recording. The result may be a significant improvement over an older, slow multiplexed metabolic recording with a cycle time of 20–60 min.

In addition, if all sensors associated with the cage are sampled at the same rate as the gas exchange data (for example 1 Hz), synchronization between cage sensor data and metabolic data is essentially perfect. The metabolic data may lag slightly behind the data from the other sensors because of plumbing considerations, but this lag (typically < 30 sec) is essentially constant at a given flow rate and is easily corrected for if needed. With cautious application of response correction (Chapter 8), a very close approximation to the second-by-second metabolic rate of the animal is possible. A typical example of continuously acquired data is shown in Figure 13.3.

It is possible to obtain truly continuous metabolic recordings without baselines by using "background baselining," which is described in Chapter 12. Ideally, two gas analysis chains should be used for each animal, as in Melanson et al. (2010), though with care this ideal can be relaxed somewhat to about (N + N/4) analysis chains, where N is the number of the animals to be measured. If implemented for that purpose, it will allow metabolic phenotyping centers to obtain uninterrupted metabolic data from many animals simultaneously. However, the benefits of this strategy are minimal (see "Multiplexed versus continuous sampling").

Multiplexed versus continuous sampling

Continuous metabolic phenotyping systems utilizing a separate gas analyzer chain for each experimental animal are available commercially. In keeping with the Red Queen race endemic to the biomedical research instrumentation field (see the end of Chapter 12, and also Chapter 16 for a discussion), be aware that because of temporal resolution limitations imposed by the cage time constant (see Chapter 8 and further discussed below), continuous metabolic phenotyping systems with a sampling interval shorter than about

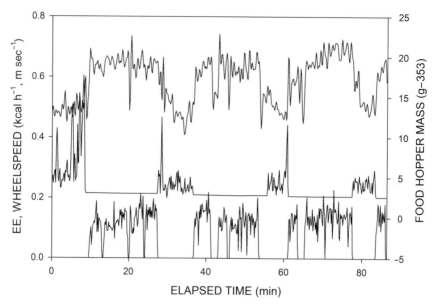

Figure 13.3 The relation between energy expenditure (EE) (top trace), raw food intake as reflected in hopper mass variations (middle trace), and wheel running speed (bottom trace). Note that the wheel speed is divided by two for scaling visibility purposes. Data are from one of eight simultaneously monitored C57BL/6J mice at 21°C, using a Promethion metabolic phenotyping system with 1 Hz temporal resolution. The energetic consequences of switching between feeding and locomotion behaviors are clearly visible. Moreover, sufficient detail is available from such recordings to extract quantitative relationships between energy expenditure and locomotion speed (see Figure 13.7).

half the duration of the cage time constant will provide negligible improvements in temporal resolution (Shannon and Weaver, 1963).

Yes, the cage time constant rears its head again (see Chapter 8). It is intuitively tempting to think of the air stream exiting a cage as reflecting the resident animal's instantaneous metabolic rate *at that moment*, but this is an ignorant fallacy. Again, most metabolic phenotyping systems use a low flow rate (~ 0.4 L min^{-1}) and a home cage with a volume of ~ 8 L, yielding a time constant of > 20 min. Thus, the cold equations of mixing kinetics (Chapter 8) ensure that the O_2 and CO_2 concentrations leaving the cage are mostly reflective of the animal's mean EE over the last ~ 20 min, mingled with the slowly fading ghosts of fossil EEs dating as far back as an hour or two. For example, a continuous system with a cage time constant of 20 min and a sampling frequency of 1 Hz is oversampling its sluggish gas concentration signals by a factor of ~ 1200 in the delusional belief that it improves the system's temporal resolution by a similar factor. If this is the goal, such a system is decoratively yet dysfunctionally burnishing a coprolite.

An obvious way to improve this situation is to increase the flow rate, thus speeding up the time constant. However, this requires low-noise gas analyzers that may not be available. That is unfortunate because the effect of the time constant on data quality is significant, especially regarding fast-changing EE signals. This can be demonstrated by modeling the effects of different time constants on a rapidly changing metabolic signal (Figure 13.4).

It is obvious from inspection of Figure 13.4 that simply sampling the bottom trace more rapidly does not improve its unacceptably poor temporal resolution. Conversely, sampling the middle trace at a relatively slow rate of every 2.5 min (as shown; easily achieved with a good multiplexed system sharing eight cages per analyzer chain) yields far superior temporal resolution compared with oversampling the bottom trace.

It is reasonable to ask whether the Z transform (response correction) as outlined in Chapter 8 can be used to mitigate the effects of the cage time constant. The answer is yes, sometimes, cautiously. It must be borne in mind that the Z transform assumes perfect convective mixing within the cage, which in a typical situation as with room

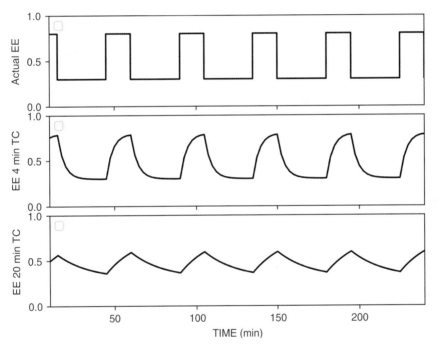

Figure 13.4 Demonstrating the effect of the cage time constant (TC) on metabolic measurement. Top trace: A hypothetical mouse alternating EE between 0.3 and 0.8 kcal h^{-1} (resting and active states, REE and AEE, respectively). Middle trace: The time response of a metabolic measurement system with a cage time constant of 4 min, sampled at 2.5-min intervals with linear interpolation between samples. REE and AEE are clearly distinguished: AEE is measured with fair accuracy, REE is measured with good accuracy. Bottom trace: As above, but with a cage time constant of 20 min, sampled every second. AEE and REE are blurred together and neither can be measured with acceptable accuracy, regardless of the rate at which the signal is sampled—which cannot change the limitations of the underlying data. Python code for this simulation is available in Appendix 3.

calorimetry (Chapter 12) requires the use of fans and vigorous stirring of the air. Unfortunately, the convection that makes the Z transform feasible in a small cage may also convectively draw heat from the animal. Because the mouse or rat is attempting to defend a core body temperature of 37°C it must increase its EE to compensate for that heat loss. It is difficult to quantify this effect in a real mouse with a highly variable metabolic rate, but by using a thermoregulated model mouse with a controlled "body temperature" of 37°C (see Bakken, 1992) the effect of vigorous convection within the confines of the cage becomes easy to see (Figure 13.5).

When the fan is activated, the EE of the model immediately increases to compensate for the elevated convective heat loss, then decreases slightly as the power dissipation of the fan elevates the temperature within the cage, reducing the thermoregulatory energy demand (in-cage temperature data not shown). When the fan turns off, convective heat loss decreases and EE plummets, then rises again slightly as the temperature of the cage falls because the fan is no longer dissipating power within the cage. The model had no fur and could not move, so these effects may be reduced in an actual animal that has control over piloerection, vasomotor tone, and posture.

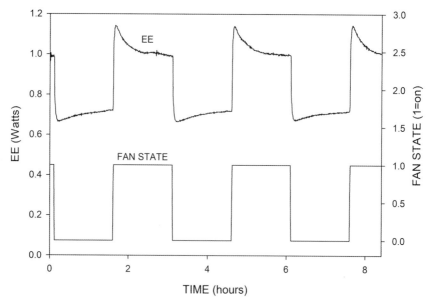

TIME (hours)

Figure 13.5 The effect of the in-cage fan on EE of a heated metal mouse model in an individual Columbus Instruments metabolic phenotyping system cage at a controlled ambient temperature of 20°C. The model was regulated at a constant 37°C by a resistive heater in a PID control loop, using a thermocouple thermometer to read its temperature. Upper trace: EE of the model, calculated from heater voltage and current. Lower trace: Fan state, 1 = on, 0 = off, in the cage lid. The fan, as supplied in the lid, is a Sunon KD1206PHB1 rated at 588 L min⁻¹ free air flow. It is turned off and on every hour in this experiment by an automated control system to investigate the metabolic effect of convective heat loss from the model. (See text; Lighton, unpublished data.)

If, as Figure 13.5 suggests, compensating for convective heat loss is a significant part of EE in cages using forced convective mixing, then one would expect that the EE of mice within such a cage to scale strongly in proportion to surface area, the sole avenue through which heat is lost by convection. This would impose a mass scaling exponent of 0.67 (M^2/M^3), similar to the "expected" Kleiber exponent (see Chapter 15)—yet irrelevant to the actual scaling of EE on body mass in the absence of convective heat loss. This may introduce a significant complication when utilizing ANCOVA models which assume that body mass itself, rather than a separate covariate of body mass, is responsible for the EE scaling effect (see Chapter 15). Stronger mass scaling of EE on body mass in convective chambers compared with chambers relying on more passive air movement would suggest that this is a real-world problem; this might be an interesting research project.

An alternative to forced in-cage convection is to sample air from the cage using a distributed manifold with hundreds of individual ingress points. This is similar to the recommended approach in room calorimetry (see Chapter 12) and helps to ensure a well-mixed sample of air from the cage without imposing additional convective heat loss. I have experimented with Z transformation of metabolic data from manifold-equipped cages with some success, but this is a work in progress.

Push versus pull mode

For multiple-animal metabolic measurement with experimental animals such as mice or rats, pull mode respirometry is preferable to push mode. This is primarily because complex cages for mice or rats with wheels and intake monitors, such as those used in most metabolic phenotyping systems, are difficult to make leak-proof in the first place, and practically impossible to maintain in that state. Moreover, push mode cages must be custom-manufactured with tight seals and will generally not resemble the cages in which the experimental animals normally live. This means that the experimental animals may require a prolonged period of acclimation to their new surroundings before their metabolic rates reach equilibrium values.

Yet, curiously, most manufacturers of metabolic measurement equipment for model animal studies persist in producing push mode systems. This is because producing multiple pushed streams of air at defined flow rates is cheaper and easier than maintaining multiple pulled air streams, each of which may require a separate pump because the excurrent air streams from the chambers cannot be intermingled prior to analysis. Moreover, either the flow rates entering the cages should be dried prior to measuring and controlling them, or the water vapor content of the air must be measured and mathematically compensated for (eq. 8.6). The general practice is to dry the air streams which exposes the animals to 0 percent relative humidity air, which desiccates their respiratory mucosa and may cause other issues. Curiously, animal care and use committees do not seem to be aware of this potential problem.

In pull mode, cages can be equipped with food uptake and water uptake monitors, food access control, and other advanced features without undue concerns about leaks, which are in fact required for the system to operate correctly. Recall from Chapter 12 that only a miniscule inward pressure differential is required to operate a pull system without significant outward leakage (Xu, 2007). The air pulled into the cages is at ambient relative humidity, so issues related to exposing the animal to desiccated air are eliminated. In addition, if a power failure occurs and the pumps shut down, sufficient gas exchange between the cage and its environment will occur to keep the experimental animal— which might be a practically irreplaceable transgenic mouse, for example—in good health, whereas in a push system it will quickly die of asphyxiation.

Incipient hypercapnia

As we have established in Chapters 9–11, the magnitude of delta gas concentrations generated by metabolic signals is inversely proportional to the rate of air flow through the animal's cage or chamber. This means that measurement resolution and temporal resolution are in fundamental conflict. To increase the delta gas concentrations, and thus maximize gas concentration resolution, flow rate must be minimized. However, to maximize temporal resolution by minimizing the time constant of the chamber (which is chamber volume divided by flow rate), we need the highest possible flow rate.

Because of this tradeoff, most metabolic phenotyping systems operate at low flow rates because they utilize gas analyzers with relatively low resolution, and their users accept low temporal resolution because the cycle time of such systems is so long that chamber time constants are relatively unimportant—plus, that's how it's always been done. The low flow rate translates in turn to a high concentration of CO_2 within the chamber, frequently reaching 0.5 percent or more. As mentioned in Chapter 12, such incipient hypercapnia can affect human cognition and behavior. Whether this is the case with model animals such as mice or rats is an open question but should be considered a possibility. Again, animal care and use committees seem unaware of this potential problem, although they are generally aware of a requirement that the full volume of the cage should be replaced a minimum of 15 times per hour. Very few metabolic phenotyping systems can meet this specification, which requires a flow rate of ~ 2 L min⁻¹ for an 8-L cage, so exemptions from Institutional Animal Care and Use Committee (IACUC) air exchange requirements are usually requested.

To reduce the likelihood of hypercapnic effects, flow rates should be maximized to the extent compatible with acceptable gas concentration resolution. When used in a metabolic phenotyping system with a fast cycle time, high flow rates operate synergistically to minimize the possible deleterious effects of hypercapnia and maximize the system's temporal resolution.

The effects of body mass and temperature on energy expenditure

The analysis of metabolic phenotyping data is still something of a minefield, and many analytical practices in the biomedical field, notably those compensating for differing body masses, are—to put it charitably—dubious. In addition, there is the matter of the ambient temperature at which measurements are made, which can have a large and confounding effect on model animal metabolic rates. Fortunately, some of the leading figures in the biomedical research field (i.e., people to whom their fellow biomedical researchers might pay attention) and some comparative physiologists are now calling for reforms and have published some excellent guidelines on correct analytical practices for scientists who use metabolic phenotyping as a research tool. These include Tschöp et al. (2012), Kaiyala and Schwartz (2011), Butler and Kozak (2010), Speakman (2013), Meyer et al. (2015), and Lighton (2017), to all of whom the reader is referred for more analytical details.

Boiled down to their essence, these and other similar papers advocate against using simple division by a body mass parameter when comparing experimental groups with different body masses; rather, they advocate using appropriate statistical methods such as analysis of covariance (ANCOVA) (see Chapter 15 for more details); and they also advocate awareness of the effects of ambient temperature on metabolic parameters (ditto). It may be that regarding the correct acquisition and interpretation of metabolic phenotyping data, the field is finally moving out of its long phase of "pluralistic ignorance" (see Prentice and Miller, 1996, and references therein).

These topics are covered in some depth in Chapter 15. I respectfully counsel anyone interested in the analysis of metabolic phenotyping data to read that chapter and the references it contains, plus the citations mentioned above, before rushing into print.

ACTIVITY QUANTIFICATION

XYZ arrays

Without quantification of activity, metabolic data are difficult to interpret. There are many ways to quantify the activity of an experimental animal, and the subject is treated in some detail in Chapter 18. In the field of metabolic phenotyping, activity quantification almost invariably takes the form of an open-field array of infrared light beams that crisscross the cage in the X and Y axes. An additional, single array at a greater elevation is often added, intended to detect rearing by the experimental animal. This is somewhat misleadingly referred to as the Z axis—misleading in the sense that it offers only a binary detection of rearing, rather than a quantification of rearing height, while the location of the rearing is determined by the X and Y axes. The term XYZ array is often used as a synonym for an open-field detection system.

The beam spacing of the XYZ array varies considerably between systems, ranging from 1 to 2.5 cm or more. Typically, the beams are scanned rapidly, at 50–400 Hz depending on the system, and any interruption of the beams (or restoration of beam transmission) is taken to denote activity by the animal. From the pattern of broken and restored beams, the position of the animal can be determined. Thus, most XYZ arrays provide at least two levels of data: animal activity, quantified as beam-breaks, and animal position, quantified as an XY coordinate plus a nonzero Z coordinate if the animal is rearing. Animal position can be further refined by calculating the centroids of the blocked beams, increasing the effective resolution of the XY array. For example, a native 1-cm beam spacing can legitimately be interpolated to a calculated centroid position within 2.5 mm. (In keeping with the Red Queen race among biomedical manufacturers, exaggerated claims of centroid precision should be taken with a mole or two of salt.) Moreover, beam-breaks can be of two distinct types, which most systems can separate: the breaking of a single beam and perhaps an adjacent beam, indicating so-called fine movement such as grooming or scratching, and the breaking of multiple beams in a short period of time, indicating so-called coarse movement such as directed locomotion. Because beam spacing differs between different manufacturers there is no real standard for activity measurement, and results between systems are not easily comparable. Thus, beam-breaks are chiefly useful within, rather than between, studies.

Beam-break arrays require careful adjustment of height, so that the beams pass above the bedding but below any in-cage feeders and other accessories. Bedding should be of the granular type rather than fluffy varieties that the mouse can pile up, thus obstructing the beams.

Most beam-break systems impose a refractory period of 200–500 msec after each beam-break at a single location is detected, to minimize the effect of meaningless transitions by a beam that happens to be partly interrupted by the animal.

It would seem logical to make beam-breaks consistent between platforms by treating the XY array data as total distance traversed by the animal, above a settable threshold of movement. If this cumulative distance is expressed in meters, then beam-break spacing will no longer be the primary determinant of activity indices. Fine movement can be quantified by setting the movement threshold to zero, and coarse movement by setting it to a higher value. An index of fine movement alone can then be obtained by subtracting the two traces. Of course, the higher the native spatial resolution of the XY array, the better.

In a system capable of recording raw data, it is possible to iteratively change the movement threshold for cumulative locomotion detection and dynamically recalculate the cumulative distance as a function of the movement threshold. I have found an interesting effect when utilizing this technique (Figure 13.6), which it is not possible to observe in systems that preprocess their data. The cumulative distance moved by the experimental animal is logarithmically related to the threshold above which movement is accumulated. This relationship is remarkably consistent across animals and would be interesting to investigate regarding treatment effects.

In most metabolic phenotyping systems, metabolic data are acquired at one interval and XYZ beam-breaks/position and other data are acquired at other intervals peculiar to that sensor. This makes synchronization difficult, as alluded to above, unless overall metrics for long periods of time, such as circadian phases, are all that are required. The greater the extent to which all metabolic and sensor information can be synchronized

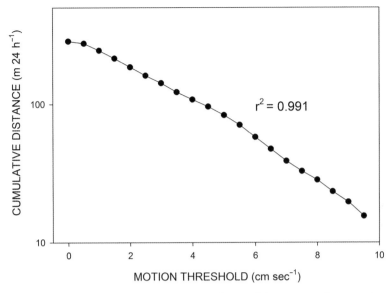

Figure 13.6 The cumulative distance in meters moved around the cage in 24 h (Y axis, logarithmic scale) above a minimum movement in centimeters per second (X axis). Coefficient of determination (r^2) = 0.991. Data are from a male C57BL/6J mouse at 21°C, using a Promethion metabolic phenotyping system. The cumulative distances were iteratively calculated in a data analysis script based on a motion threshold that increments from 0 to 9.5 cm sec^{-1}.

within a single file at a single sampling interval, the better. This makes metabolic interpolation (Figure 13.2) particularly useful.

Running wheels

Voluntary running wheels constitute an alternative method for quantifying directed pedestrian locomotion. The number of running wheel revolutions is collected in time bins, the duration of which can vary from a second to many hours depending on the system. Distance run is calculated by integrating the wheel rotation counts against time and multiplying by the circumference of the wheel. Provided the binning interval is short enough, it is also possible to derive running speed information by differentiating the summed distance data against time.

As anyone who has watched a mouse using a running wheel will realize, wheel-running is by no means a continuous activity. Typically, bouts of wheel-running may last anything from a few seconds to a few minutes, with frequent breaks for other activities. Utilizing the shortest feasible bin duration for summing wheel rotations allows the true complexity of wheel-running behavior to be quantified. It is worth noting that coarse bin intervals might not be able to differentiate between two treatment groups, one of which runs for a minute every other minute over the course of an hour, while the other runs

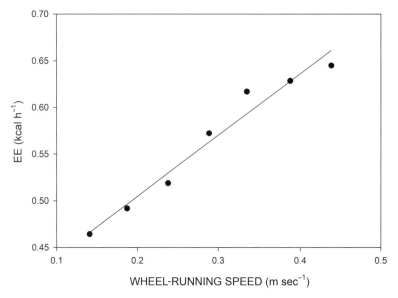

Figure 13.7 The relation between binned voluntary wheel-running speed (X axis) and energy expenditure (Y axis) of a single, voluntarily running male C57BL/6J mouse, body mass 27.7 g, at 21°C. EE and running speed data were binned at 0.04 m sec^{-1} intervals. The slope of the line is known as minimum cost of transport (MCOT). In this case it corresponds to 29 J kg^{-1} m^{-1}, close to the predicted value based on treadmill running for an animal of this body mass (Taylor et al., 1982). Coefficient of determination (r^2) = 0.962. Data are from a Promethion metabolic phenotyping system monitoring eight mice simultaneously.

continuously for 30 min over the same time interval. As usual, too much data are better than too little; you can always ignore what you don't need, but can't go back in time to acquire what you didn't get. If you can do that, why are you reading this?

Because running-wheel activity is entirely voluntary, it can under some circumstances act as a quantified exercise correlated with metabolic parameters. Other techniques for quantifying exercise in small model animals generally require the use of treadmills, which are stressful to the animals (electric shocks are often required to ensure that running is maintained) and require training and skill to operate. Using a suitably configured metabolic phenotyping system with sufficient temporal resolution (i.e., a fast time constant; see Chapter 8), it is possible to obtain good correlations between voluntary wheel-running speed and energy expenditure (see Figure 13.3) similar to those obtained using a dedicated treadmill (Figure 13.7) but without the stress, animal and operator fatigue, added complexity, and expense.

Food and water intake

Food and water intake are generally measured gravimetrically. The mass of a food hopper or water dispenser is measured, and the disturbance caused by intake activity serves

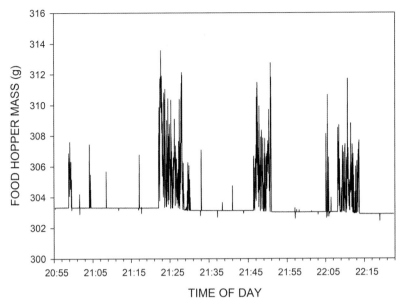

Figure 13.8 Six food intake bouts: three large and three small. Starting at 20:58, duration 1.00 min, intake of 7.9 mg; at 21:04, 0.62 min, 4.6 mg; at 21:22, 8.38 min, 161.7 mg; at 21:32, 0.40 min, 7.1 mg; at 21:46, 4.78 min, 96.9 mg; and at 22:08, 5.72 min, 129.8 mg. Mass disruptions caused by mouse interactions with the food hopper are clearly seen, from which a force application profile or force–time integral of the intake bout can be derived if desired. All of the above intake bouts were automatically validated for statistical significance using Student's t test of pre- versus post-bout hopper masses at P < 0.001. The SD of all intake events was <= 0.7 mg. Several interactions with the hopper did not result in significant intake. Data were acquired digitally from a load-cell-based sensor (Promethion MM-1), sampled at 1 Hz and recorded in raw form as shown. Data are from a male C57BL/6J mouse at 21°C, one of eight monitored simultaneously by a Promethion metabolic phenotyping system.

as a signal that an intake bout is taking place. The beginning and end of the intake bout are recorded, and the stable mass of the food hopper after the end of the intake bout is subtracted from the hopper mass before the bout began (Figure 13.8). Depending on the sophistication of the system, only the difference may be stored (and optionally rejected if it is negative), or the two masses can be more rigorously compared using a statistical test and accepted as a legitimate intake bout if the probability of the mass change occurring by chance falls below a settable threshold.

The mechanics of mass measurement differ between systems. In some systems, the hopper is placed above a mass-measuring device; in others, it is suspended from one. Some systems utilize isometric force transducers (which physiologists will recognize as the type used, for example, for gastrocnemius muscle twitch experiments), while others utilize load cells of the type used in laboratory balances, with the latter generally producing better results with less drift and less frequent recalibration requirements. Either transducer style is difficult to render leak-proof in a push mode system. The electronics also differ between systems, especially regarding the analog to digital converter resolution. Most intake systems can resolve to about 13–14 bits in practice, or roughly one part in 8000–16,000. This means that if a food hopper can weigh up to 160 g, the system can resolve a mass change no smaller than 0.01 g, or, in practice, about 0.02 g. Other intake systems utilize more sophisticated technology with digital data transfer and can resolve to approximately 1–2 mg over a 1-kg range, or one part in 500,000. Such systems can resolve smaller intake events than other systems can detect. For example, the three smallest intake events shown in Figure 13.8 are below the limit of detection for most metabolic phenotyping systems, but are easily quantified with negligible error by a higher resolution system. Such "micro-intake" events, ranging from ~ 3 to 20 mg, constitute ~ 30 percent of all food intake events (author's unpublished observations; see Figure 13.9). They may be negligible compared to a total food intake budget, but each such event is prompted by neurophysiological phenomena prompting feeding behavior followed by rapid satiety that may be of interest to the researcher. Again, it is better to record and discard such events if they are not relevant to your research than to remain unaware that they occurred in the first place. The choice of which data to use or ignore should be yours, not that of your metabolic phenotyping system.

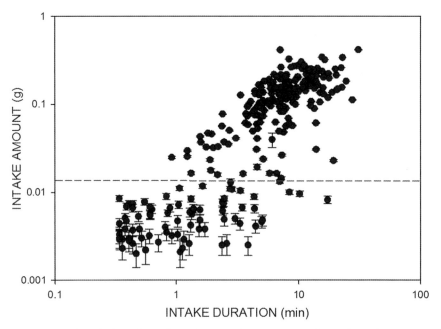

Figure 13.9 Food intake bouts by eight male C57BL/6J mice over 24 h at 23 °C ambient temperature. The error bars denote standard errors of the pre- versus post-intake hopper masses. Note that the ordinate and abscissa are logarithmic. The dashed line represents the approximate lowest threshold of detection for intake events by the majority of metabolic phenotyping systems in common use. Data are from a Promethion metabolic phenotyping system.

Systems also vary in the amount of information they obtain from intake phenomena. Some will totalize intake measurements; others will, in addition, provide information on the duration and intervals between intake bouts. Yet others will record all the data from the mass measurement sensors agnostically, providing researchers with intake data that can be mined in any way they choose.

Most systems allow the researcher to define what constitutes a single intake bout. Many intake bouts consist of several discrete intake events separated from each other by pauses of a few seconds or tens of seconds. The researcher should thus decide on the maximum pause duration allowable within a single intake bout before it is counted as two separate bouts. Similarly, if a researcher is interested in quantifying meals, he/she needs to define the maximum interval between individual intake bouts within a meal. Both intake bouts and meals can usually be associated with a minimum intake amount below which the event is ignored. Or, in the case of a meal, it may be considered as a series of discrete intake events that does not merit status as an actual meal.

All the above are applicable to water intake as well, which in some systems is co-quantified with food intake as a meal. For an excellent review of feeding behavior see Ellacott et al. (2010).

Food contains some adsorbed water, about 10–15 percent for typical rodent chow. The degree of chow hydration is determined by the chemical makeup of the chow and by its free energy for absorption, which is proportional to the ratio of actual water vapor partial pressure to saturated water vapor at the temperature of the chow; this is in turn equivalent to water vapor activity, or RH%/100. Thus, changes in intra-cage RH% cause variations in hydration level and thus food mass. This need not impair the accuracy of intake mass measurements, which are inherently differential in most food intake measurement systems (see Figure 13.8), so slow baseline shifts are largely canceled out. However, if energy equivalents are needed, varying hydration levels add significant noise to energy conversion factors. Dynamically calculating conversion factors determined from empirical relations between intra-cage relative humidity and chow hydration levels may reduce this source of inaccuracy, provided that reliable intra-cage RH% values are available, as in systems that utilize mathematical compensation for water vapor dilution of respiratory gases (see Chapter 8).

BODY MASS

The regular weighing of mice or rats required in many experimental protocols is burdensome for humans and stressful for the animals being weighed. By suspending an enrichment habitat attractive to the animal (e.g., made from dark red polycarbonate that appears opaque to it) from a suitable mass monitor, it is possible to weigh the animal accurately each time it enters the habitat and each time it leaves it. This can yield a fine-grained, high-resolution record of body mass over time without stressing the animal or inconveniencing the researcher. If raw data are retained, additional information such as

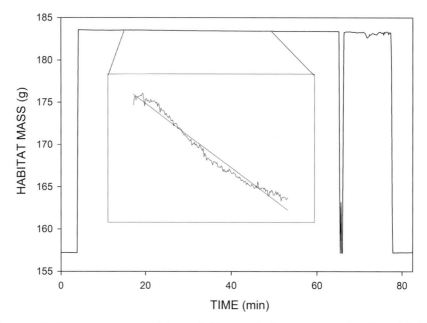

Figure 13.10 A mouse enters an enrichment habitat attached to a mass monitor, leaves it briefly, then enters and leaves it again. During the first event, it remained quiet and lost mass because of water evaporation and CO_2 emission at 0.252 g h^{-1} (linear regression analysis, r^2 = 0.925 during the indicated section; inset shows a zoomed view). During the second event, it groomed itself, visible as a disturbance of the mass trace. The body mass habitat can therefore also serve as a sensitive activity sensor (see Figure 13.12 for an application of this). Mouse mass was 26.41 g at start. Data were sampled at 1 Hz from a male C57BL/6J mouse, ambient temperature 21°C, using a Promethion metabolic phenotyping system.

the occurrence of grooming episodes can be coaxed from the data (Figure 13.10). It is also possible, from the slope of body mass versus time within a habitat occupation, to measure water loss rate from the experimental animal, though this can also be accomplished with a gas analysis system that includes water vapor detection.

A wealth of information is therefore available both from the raw data (e.g., duration of habitat visits, grooming behavior, and mass balance during each visit) and by processing the raw data to reveal overall patterns of mass balance across the duration of the recording. In this respect it should be noted that dynamic recording of experimental animal body mass is an important but neglected area of metabolic phenotyping. It is a useful indicator of an animal's stress status; if stressed, it will tend to lose body mass, which has been regarded as a characteristic phenomenon in most metabolic phenotyping systems. An example of dynamic tracking of mouse body mass is shown in Figure 13.11.

In addition, accurate measurement of RQ, especially if it is being compared to the food quotient (FQ) of the animal's diet, requires knowledge of body mass balance. If an experimental animal gains or loses body mass, its RQ will be greater than or less than the

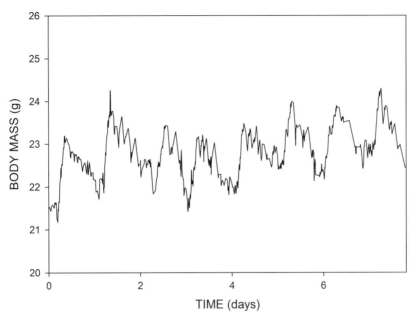

Figure 13.11 Dynamic tracking of the body mass of a male C57BL/6J mouse at 21°C. Each reading corresponds to an entry into or exit from a body mass habitat. Individual data points are joined by linear interpolation. Circadian cycles in body mass balance are clearly visible. Over the ~ 8-day duration of this recording, starting immediately after being introduced to the cage from a communal Jackson Labs shipping carton, the mouse gained body mass at a rate of 0.107 g day^{-1}. These unpublished data are from a Promethion metabolic phenotyping system.

FQ of its diet (Longo et al., 2010). By the same token, the RQ of an experimental animal will be equal to its FQ if it is in body mass equilibrium. This is a useful though neglected method for in vivo, in situ validation of a metabolic phenotyping system. It is disturbing to note the number of published investigations in which the FQ of an animal's diet is quite different from its 24-h RQ (author's observations and O. McGuinness, personal communication). I deal with the FQ = RQ method for validating system accuracy in Chapter 14, and (somewhat seditiously) with the role of RQ in calculating the energy equivalent of gas exchange data at the end of Chapter 9.

Automated weighing does have some disadvantages. Primarily, the researcher is at the mercy of the experimental animal's proclivity to enter its enrichment habitat. If it does not enter the habitat, it does not get weighed. (In my experience, this seldom happens.) Specially commodious habitats may be necessary for gravitationally challenged animals such as ob/ob mice. For larger animals with long tails such as rats, the tail may touch the cage bedding while the animal is in the habitat, yielding a noisy and slightly underestimated body mass record. Designing an enrichment habitat that encourages or even requires the rat to curl its tail into the habitat may address this problem, which to my knowledge does not yet have a satisfactory solution.

ENVIRONMENTAL PARAMETERS

Because the importance of environmental variables such as temperature is now recognized in the metabolic phenotyping field, it is important that they be measured and reported. At a minimum, ambient temperature and relative humidity should be recorded. Some metabolic phenotyping systems record water vapor as part of their data stream, both within and outside the cages, so a separate measure of relative humidity may not be required.

Experimental animals are sensitive to noises and to the presence of humans, and thus a complete environmental monitoring system should include occupancy detection and at least a quantitative measure of ambient sound levels. Light level is an exceedingly important parameter and should be recorded as well. Light level data can act as a useful check for the start and end of circadian phases and is especially useful if circadian cycle length is manipulated so that its relationship with clock time is complex or unpredictable. Occupancy, sound level, and light level can all be qualitatively measured unless quantitative data (e.g., light level in Lux) is explicitly required, which is seldom the case.

On a more speculative note, some small animals are known to detect variations in barometric pressure as an index of changes in weather, and may alter their behavior accordingly (Breuner et al., 2013). If barometric pressure data are available (they are a standard part of the data stream in systems that utilize mathematical compensation for water vapor dilution; see Chapter 8), they should also be recorded.

As in everything having to do with research, the golden rule is to eschew paucity and embrace the richness and variety of your data to the maximum extent possible. This may require some re-education. Storage and CPU cycles should be regarded as infinite because the arc of progress strongly bends that way. It is far easier to cull a rich data set and ignore items seemingly not relevant to your research than to lack a critically important stream of data, only realized as such in retrospect, which could have elevated your research project from mundane to truly great.

TRACEABILITY

Most of the metabolic phenotyping systems in common use in the biomedical field calculate gas exchange parameters on the fly and preserve only final data calculated in a "black box" fashion; other systems coming into wider use retain all the raw time-series data from all instruments and process data after acquisition for maximum accuracy, versatility, and traceability.

There is something to be said for the on-the-fly approach. It is largely set-and-forget and can be simple to set up and maintain. Because the data are extracted in real time and the fine-grained raw data are discarded, the storage and computational requirements of on-the-fly systems are minimal, and they can continue to record indefinitely.

By contrast, the more traceable approach generates a huge amount of data, recording data from over 300 channels once per second—about 100 MB day^{-1} for a 16-cage system with intake and activity monitoring. This makes it difficult to record and analyze continuous sequences of data that last longer than about 1–3 weeks with typical computers. This will, however, change as technology catches up. Physicists and astronomers deal with far larger data sets as a matter of course and it is only a matter of time before their methodologies seep into metabolic phenotyping. The most advanced systems are already capable of secure web-based monitoring and control, and cloud-based data archiving and analysis of large data sets will probably be widespread soon after this book is published.

To balance the sheer volume of data, the traceable raw data approach allows the listing and *understanding* of every step from the raw data to the extracted, final data. If necessary, the data can be reanalyzed in the light of later knowledge and improvements in analytical techniques. For example, every reading extracted from a metabolic phenotyping record that records all raw data can be traced back to the original data from which it was derived (e.g., Figure 13.2), without which meaningful quality control is difficult or impossible. This also means that if, for example, a new algorithm is developed for meal pattern analysis, it can easily be performed on data that have already been acquired. If data need to be binned in a different way, the same applies; rerunning the experiment is not required, minimizing time, institutional resources, and experimental animal usage. Different laboratories can be assured of compatible data if they analyze their data using shared analytical scripts. By using Internet-enabled systems, distributing research projects across multiple laboratories becomes practical. Balancing these advantages is the fact that investigators accustomed to black-box systems may find this versatility unnerving. The on-the-fly systems use internal algorithms that are opaque and described, if at all, only in generalities that are impossible to verify because the data from which the results were derived have long since joined the choir invisible. Therefore, the extracted data may acquire an air of magisterial finality that in my opinion is both spurious and misleading, but to which many workers in the biomedical field have become accustomed. Convincing people who are "assured of certain certainties" (Eliot, 1917) to adopt a different approach is uphill work, but necessary if metabolic phenotyping is to progress and become compatible with good laboratory practice in other fields.

As an example of the benefits of raw data storage, I can offer a personal anecdote. Exploring a data set acquired at 1 Hz from a mouse that alternated between a running wheel, a food intake monitor, a water intake monitor, and a body mass monitor, and graphing the outputs of all relevant sensors at once together with position in the cage, I was forcibly struck by the fact that I was looking at a graphic display of sequential mouse behaviors. Each behavior could be placed precisely in time and rigorously quantified. This is equivalent to focal animal behavior monitoring (Altmann, 1974)—but automated, and for all animals simultaneously. Writing code (which I have called EthoScan) to extract time budgets, locomotory budgets (both wheel and nonwheel, the latter from the XY position sensors), and lists of behaviors with their associated numeric correlates (time of initiation, duration, distance traveled during the behavior, and a quantitative index such as grams of food eaten, milliliters of water drunk, body mass, etc., dependent on the specific behavior) was then straightforward. Moreover, from the behavioral sequence it is possible to generate a transition probability matrix that expresses the probability that one behavior will be followed by another, and which therefore provides a fascinating window into the way that (so to speak) the mouse's brain is wired, especially with visual aids such as Markov chain visualization (Lark et al., 2018 shows an excellent use of this approach, together with hierarchical object clustering of behavioral data). None of this was imagined before those data were acquired—yet the new analytical procedures do not care, and work with compatible data of any age. The EthoScan technique approaches the behavioral data richness of video analysis, but with greater quantification (e.g., exact food intake figures), less fuss and lower expense. None of this would have been possible, or indeed foreseeable, without retention of all raw data. This only scratches the surface of what is possible if all raw data are retained.

COMBINING BEHAVIOR AND METABOLIC DATA

As my mentor George Bartholomew once wrote, "If physiology is defined as the study of vital functions, it becomes inseparable from morphology and behavior" (Bartholomew, 1958). Studying physiology and behavior in isolation is absurd unless only the most reductionist questions are being asked, yet many scientists regard them as two entirely separate disciplines. This need not be the case.

I will content myself with just one simple but speculative example derived from work in progress, in collaboration with Karl Kaiyala of the University of Washington. On scanning EE data while mice were in their body mass habitats for periods of at least an hour, I had the initial goal of determining EE during sleep. This is because body mass habitats are exquisitely sensitive to any movement by the mouse (Figure 13.10), so periods of guaranteed inactivity in the habitat could be determined. I then discovered to my surprise that about 20 min prior to emerging from the habitat, while a mouse is still completely motionless, its EE rises to a level very close to its normal active value, whereupon it becomes active and leaves its habitat. Moreover, its RQ starts to rise at the same time as its EE, suggesting that its motionless EE elevation is fueled by a shift in metabolic substrate towards carbohydrates. This was consistent across 15 such arousals (Figure 13.12).

What could be the cause of this? Any form of crossbridge cycling, including bouts of shivering, would have left obvious evidence in the form of increased sample-to-sample mass variance. Brown adipose tissue (BAT) activation supplied a possible answer, but BAT would normally be expected to oxidize lipids, causing a decrease rather than an increase in RQ. When two colleagues and I presented a poster reporting this finding at a symposium, a well-known metabolic expert dismissed it as an instrument malfunction. I later sent a copy of the averaged 15-event graph shown in Figure 13.12 to the expert and asked him to speculate on the malfunction that could explain the data but received no

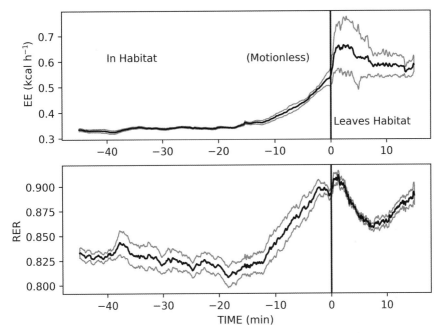

Figure 13.12 While completely motionless in a habitat used in conjunction with a mass sensor to measure body mass, the EE of a C57BL/6J mouse at 21°C begins to increase about 20 min prior to emerging from the habitat. Motionlessness during the period of EE elevation is confirmed by negligible variation in sample-to-sample body mass during this period (see Figure 13.10). Starting at the same time, the RQ of the mouse increases, indicating a shift in metabolic substrate utilization towards carbohydrates. Mean of 15 arousal episodes synchronized to the moment of exiting the habitat (time = 0); gray lines are ± 1 SE. (See text; Lighton and Kaiyala, unpublished data.) RER, Respiratory exchange ratio.

reply. Interestingly, a recent study (Held et al., 2018) demonstrated in vitro that BAT can and does oxidize glucose. Figure 13.12 suggests, using a combination of behavior and metabolic data, that this may occur in vivo as well.

As integration of metabolic physiology and behavior becomes more routine and the richness of data sets continues to increase (e.g., Lark et al., 2018), there will be many more examples to follow.

14

Validating Flow-through Respirometry

Validating the results obtained from flow-through respirometry systems is an important fault-finding, or even confidence-building, exercise. Although several ingenious and even bizarre techniques have been described, I concentrate here on three sensible and approachable methods: nitrogen injection, the alcohol lamp, and a gas flame.

NITROGEN INJECTION

Nitrogen injection takes advantage of the fact that oxygen consumption can't be distinguished from oxygen dilution, provided that nothing else in the setup changes. Typically, nitrogen at a known flow rate is injected into the inlet of a system without affecting the flow rate to the system. This is easily accomplished, for example, by leading a short section of tubing into the incurrent air connection of a respirometry system and leaving plenty of space for the main flow to enter. The nitrogen can then be injected into the system at any time. Of course, though nitrogen is an economical and traditional choice, you can use any other gas that will not change the characteristics of the system, such as argon, so this is a good excuse to buy a cylinder of argon, which is a highly entertaining element if you're interested in high-voltage experiments.

I will keep this simple first and look at the case of injecting dry nitrogen into a flow rate (FR) of dry air. As Bakken (1991) elegantly showed, the following equation holds:

$$VO_2 = FRA\left\{F_iO_2 / \left[1 - F_iO_2 + (FRA / FR)\right]\right\} \qquad (14.1)$$

where FRA is the flow rate of dry argon or nitrogen entering the system and FR is the flow rate of dry air entering the system (note that the addition of FRA at the inlet does not change FR, which is regulated); VO_2, as always, is in the same units as FR and FRA. Obviously FR and FRA should be the same type of flow; do not mix mass and volumetric flows or mass flows calibrated at different reference temperatures without interconverting them. Furthermore, the incurrent flow rate (FR) and FRA should be accurately known. Techniques for measuring and calibrating flow rates are described in Chapter 17. It should go without saying that the O_2 analyzer should first be carefully calibrated, preferably both at zero percent O_2 and dry ambient air (20.94 percent O_2).

Measuring Metabolic Rates: A Manual for Scientists. Second Edition. John R. B. Lighton, Oxford University Press (2019).
© John R. B. Lighton (2019). DOI: 10.1093/oso/9780198830399.001.0001

If you want to simulate a given VO_2, use this equation (Bakken, 1991) to obtain the flow rate of argon or nitrogen required:

$$FRA = \{VO_2\left[FR\left(1-F_iO_2\right)\right]\}/\left[F_iO_2\left(FR-VO_2\right)\right] \qquad (14.2)$$

First calculate FRA for a VO_2 that is within the range of your experimental organism. Flow dry air through the system at known, validated FR. Then, after recording for a few minutes without adding argon, proceed to add it at a known and verified rate until equilibrium is reached. After a short while, shut it off, allow the system to return to baseline ($= F_iO_2$) levels for a few minutes, then subtract the baselines (with drift correction if possible), flip the trace, and calculate VO_2 as shown above. Do several trials to determine the statistical spread of your results. They should not differ from the predicted VO_2 if you use a Student's t-test of the collected data against a single sample of the predicted VO_2. It is worth noting that this technique is excellent for determining the response characteristics of a system when implementing the instantaneous transform (see Chapter 8).

The situation is a little more complicated if you use a pull system with excurrent flow measurement that uses ambient air as the baseline (see Chapter 11). This is because you are adding dry argon (or nitrogen) to air that contains water vapor, thus introducing an additional variable into the mix. Fedak et al. (1981) described an approach that allows a respirometry system using a high flow rate to be calibrated by a low, known flow of nitrogen, and their paper contains a fairly thorough treatment of the subject.

Here, we will assume that the main flow rate is known (see Chapter 17) and that you are measuring excurrent water vapor and drying the air before O_2 analysis. The excurrent water vapor will decline when the nitrogen (or argon) is added by the factor (FR − FRA)/FR. Because you are correcting the flow rate for the presence of water vapor (equation 8.6), this effect is automatically compensated for. If, in contrast, you are measuring ambient water vapor levels, either directly or by calculation from relative humidity and ambient temperature (equation 16.7), then you will need to multiply that measurement by the above factor while nitrogen (or argon) is being injected. As described in Chapter 4, an alternative approach is to humidify the diluent so that it approximates ambient water vapor pressure. One rather costly approach would be to flow it through a section of parallel Nafion tubing (see Chapter 19) long enough to ensure that it reaches water vapor equilibrium with ambient air.

ALCOHOL LAMP

The alcohol lamp, a venerable technique, is based on the fact that 1 mol ethanol (C_2H_5OH; molar mass 46.07 g), when burned in O_2, consumes 3 mol O_2 and produces 2 mol CO_2 and 3 mol water, the latter in the form of water vapor. Thus, 1 g of ethanol consumes (3 × 22.414)/46.07, or 1.460 L O_2 at STP. It also produces 0.973 L CO_2 and 1.460 L water vapor.

Its heat equivalent is 26.71 kJ g^{-1} or 6.383 kcal g^{-1}. As an alternative that is less expensive and less likely to disappear inexplicably, 2 mol methanol (CH_3OH; molar mass 32.04 g) consume 3 mol O_2 and produce 2 mol CO_2 and 4 mol water. Thus, 1 g of methanol consumes $(3 \times 22.414)/64.08$, or 1.049 L O_2 at STP. It also produces 0.700 L CO_2 and 1.399 L water vapor.

In theory, simply obtain a good alcohol lamp with a small wick that burns absolute ethanol or methanol with a tiny, almost invisible flame. Then (1) calibrate your O_2 analyzer, (2) record from your respirometry system with nothing in the chamber, (3) weigh the alcohol lamp, (4) immediately place it in the chamber, (5) allow the system to come to equilibrium for a reasonable period, and (6) remove the alcohol lamp and reweigh it immediately. Calculate VO_2 as usual (if you have to assume an RQ, use the figure of 2/3 or 0.667). Calculate VCO_2 and VH_2O (if applicable). Correct these figures to STP if you need to (e.g., if for some reason you are using a rotameter for measuring flow rates). Predict VO_2, VCO_2, and VH_2O by multiplying the grams of ethanol burned by the O_2, CO_2, and H_2O liters or milliliters listed above, and divide by time in the units of measured VO_2. Experimental and predicted values should not differ significantly.

If you can manage to weigh the lamp continuously during the experiment, so much the better. This can be done by using a digital scale with a serial output and incorporating the serial data stream into the data acquisition system. Alternatively, manually note values at known intervals during the recording and calculate the burn rate, which should be very constant, by linear regression. You can also use a strain gauge sensor, load cell, or linear variable displacement transducer (LVDT) plus spring to measure the lamp's mass in real time with your data acquisition system if you have the technical expertise to do so.

A minor problem with the alcohol burn technique is the fact that low metabolic rates are very difficult to simulate; a major problem is evaporation. Not all of the ethanol is burned. In practice, evaporation is almost impossible to avoid, which causes the readings to be slightly lower than they should be (considerably so, if the wick was large). Minimizing the evaporation problem requires ingenious lamp designs, such as the one described by Depocas and Hart (1957). Evaporation disturbs the absolute values for VO_2, VCO_2, and VH_2O, but fortunately not their ratios. In summary, the alcohol lamp validation method is useful but needs to be implemented with great care if it is to give useful absolute results. However, as a check on the ability of a respirometry system to measure RQ it has much to commend it.

The alcohol lamp approach can, of course, also be used with other combustible substances. A good example is the endearingly named "artificial rat" of MacKay et al. (1991), which uses a custom butane burner.

Note that if you use the alcohol lamp technique, you will have to use the correct respirometry calculations (see Chapters 9–11) and observe the equations' assumptions, such as analysis of dried and/or CO_2-free air. As a shortcut, if you are interested only in ratios, see the section "Gas flames (stoichiometric method)."

PROPANE FLAME

Propane is an excellent choice for respirometry validation, especially for larger organisms. It is the *de facto* standard in room calorimetry (see Chapter 12). One mole of propane (C_3H_8; molar mass 44.06 g), when burned in O_2, has the stoichiometric equation

$$C_3H_8 + 5\,O_2 \rightarrow 3\,CO_2 + 4\,H_2O \qquad (14.3)$$

Thus, 1 mol propane consumes 5 mol O_2 and produces 3 mol CO_2 and 4 mol water, the latter in the form of water vapor. Thus, 1 g of propane consumes (5×22.414)/44.06, or 2.543 L O_2 at STP. It also produces 1.526 L CO_2 and has a heat equivalent of 50.34 kJ g^{-1} or 12.03 kcal g^{-1}. Its equivalent RQ is 0.600 (3/5). Propane can be used for a fully quantitative burn, in which its mass loss during combustion is converted into the appropriate units and compared against values measured by indirect calorimetry. It is also useful for checking the overall performance of a respirometry system that measures animals too small to be imitated with a practical propane flame, such as mice. In such applications, an aliquot of air collected above the flame can be cooled with a copper coil and pumped at a low rate, sufficient to imitate a VO_2 of about 1–3 mL min^{-1}, into the respirometry cage being tested. The measured RQ should not differ significantly from 0.60, provided that the O_2 and CO_2 analyzers are working correctly and have been properly calibrated, and provided that the propane is pure and that combustion took place with complete efficiency.

These are quite significant assumptions. Although propane burns usually give excellent results, it should be borne in mind that (1) the propane flame should be blue and almost invisible (yellow coloration denotes incomplete combustion), and (2) not all propane is necessarily equal. Although "propane" in conventional propane cylinders is at least 90 percent propane, it may also contain ethane, butane, propene, butane, and alarm-odorants such as ethyl mercaptan. This is according to the "HD-5" standard (as defined by ASTM Standard 1835: Heavy Duty—5 percent maximum allowable propene [C_3H_6, RQ = 0.667] content, with up to but not exceeding 5 percent each of butane [C_4H_{10}, RQ = 0.615] and ethane [C_2H_6, RQ = 0.571]). The formulae of these possible contaminants are similar enough to that of propane and their concentration low enough that their practical effect will generally be minor, especially when you consider that most CO_2 span gases are accurate to only 2 percent or 5 percent of their stated value. *Caveat incendor.*

GAS FLAMES (STOICHIOMETRIC METHOD)

Oxygen analyzers are relatively easy to calibrate reliably because most suffer from very little drift at the zero point, where they are easily calibrated with nitrogen, and dried ambient air (especially outdoors or in a well-ventilated environment) will be within a very small margin of 20.94 percent (Tohjima et al., 2005). CO_2 analyzers are not so simple to calibrate, requiring a zero gas such as CO_2-scrubbed air (see Chapter 19) or nitrogen, plus a span gas of known concentration. CO_2 span gases are expensive if their accuracy is good, are difficult to travel with, and are prone to being left on by mistake and thereby being wasted. The same, *mutatis mutandis*, goes for water vapor analyzers, except for dewpoint analyzers, which are based on a primary physical property of water vapor. These factors make CO_2 and water vapor analyzers more difficult to calibrate reliably than O_2 analyzers and thus more likely to limit the accuracy of a flow-through respirometry system.

Withers (2001) makes a good recommendation: Given that O_2 analyzers are easy to calibrate, why not use the combustion of substances with a known stoichiometry to calibrate CO_2 and H_2O vapor analyzers? For methane, propane, butane, and ethanol, the stoichiometric ratio of CO_2 production to O_2 consumption is, respectively, 0.500, 0.600, 0.615, and 0.667. For water vapor, the production ratios per unit O_2 consumed are 1.000, 0.800, 0.769, and 1.000. Thus, a small, clean-burning flame in a respirometry system can be used to check the calibration of a CO_2 or water vapor analyzer by zeroing the analyzer, then adjusting its span so that it yields the appropriate ratio against O_2 consumption, using the appropriate respirometry equations.

A small shortcut: If only $F_eCO_2 - F_iCO_2$ or $F_eH_2O - F_iH_2O$ are measured and divided by $F_iO_2 - F_eO_2$, the ratios will differ from their stoichiometric ratios because of dilution and concentration effects that were explained in Chapter 9. These ratios can be used for short-cut calibrations with suitable compensation (Withers, 2001). For CO_2 versus O_2, the compensated ratios are 0.559, 0.655, 0.669, and 0.720 for methane, propane, butane, and ethanol, respectively; in the same order, for water vapor (expressed as fractional concentration or kilopascals of water vapor pressure per kilopascals of barometric pressure), they are 0.905, 0.738, 0.712, and 0.905.

FOOD QUOTIENT AND RESPIRATORY QUOTIENT EQUIVALENCE

As mentioned near the start of Chapter 6, Hess's law of constant heat summation states that the total energy released in a chemical reaction is independent of the number of steps in that reaction. This is the foundation on which indirect calorimetry rests. It follows because of Hess's law that the ratios of CO_2 produced:O_2 consumed by combusting food in pure oxygen, or by combining the food with oxygen in the body of an animal, are equivalent. The latter measurement, discussed extensively throughout this book, is of course the RQ. The former is referred to as the food quotient (FQ). It should immediately be emphasized that FQ = RQ holds true if, and only if, the animal is in body mass equilibrium. Thus, this technique should be used only if the body mass of the animal is measured periodically.

The FQ of a given food can be determined experimentally by combusting it in a bomb calorimeter (see Chapter 6), or by a simple calculation if the caloric makeup of the food is known. In the case of diets fed to laboratory animals, their compositions are generally very stable and well characterized, and can be downloaded from the Internet. Thus, for example, the commonly used LabDiet 5001 information sheet (simply Google "LabDiet 5001 pdf" to obtain it) states that the calories provided by protein are 29.829 percent; by fat, 13.427 percent; and by carbohydrates, 56.744 percent. Each of these macronutrients has a characteristic RQ. That of fat is generally considered to be 0.71, that of carbohydrate is 1.0, and that of protein will be close to 0.83, with some wiggle room owing to variable amino acid proportions. The FQ of a diet of known caloric proportions is therefore, to a good approximation, simply:

$$FQ = (F/100 \times 0.71) + (P/100 \times 0.83) + (C/100 \times 1.0) \qquad (14.4)$$

Where F, P, and C are the percentage of total calories provided by fat, protein and carbohydrates, respectively. Thus, in the case of the LabDiet 5001 rodent chow, FQ = $(0.13427 \times 0.71) + (0.29829 \times 0.83) + (0.56744 \times 1.0) = 0.911$. Consequently, if (for example)

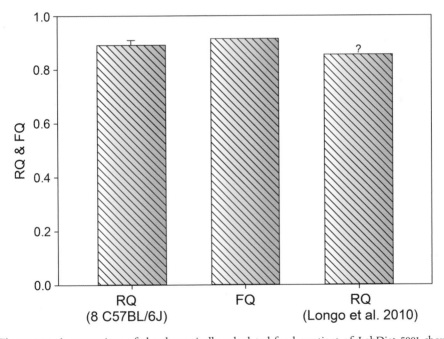

Figure 14.1 A comparison of the theoretically calculated food quotient of LabDiet 5001 chow (center bar, FQ = 0.911) and the mean RQ of eight C57BL/6J mice (left bar, error bar = 1 SD) across a 21-day period, measured with a Promethion metabolic phenotyping system at temperatures ranging from 19 to 29°C (Kaiyala, Wisse, Lighton, and Förster, unpublished data). The experimentally determined RQ does not differ significantly from the calculated FQ of the diet (P > 0.25). The right bar shows the experimentally determined RQ measured by Longo et al. (2010) on the same diet; no variance statistics were provided, so its statistical significance is uncertain.

a mouse is consuming this diet and is in body mass equilibrium, its 24-h mean RQ should not differ significantly from 0.911. This allows you to perform a useful noninvasive check on the basic accuracy of your metabolic measurement system. If FQ = RQ, provided that body mass is in equilibrium, then both your O_2 and CO_2 analysis systems are functioning optimally. This does not necessarily mean that they are delivering accurate readings of VO_2 and VCO_2—simply that the two gas analyzers are accurately measuring O_2 and CO_2 concentrations. VO_2 and VCO_2 accuracy also requires, in addition to accurate gas analysis, accurate flow measurement and/or control. This can be carried out by any of the validation techniques based on gas mixing described above. If accurate flow measurement is verified and if FQ = RQ at body mass equilibrium, then it is practically certain that your metabolic measurements are accurate.

In practice, body mass seldom maintains perfect equilibrium. Provided that a time series of body mass measurements is available, however, RQ readings can be corrected to a mass equilibrium value. By regression analysis, obtain the RQ intercept at the point where body mass change = 0 (Longo et al., 2010; S. Virtue, personal communication). Figure 14.1 shows RQ data from mice fed on LabDiet 5001 chow, and it is evident that in body mass equilibrium, their FQ and their RQ are equivalent within experimental error.

15

Data Analysis and Presentation

Any kind of respirometry, and flow-through respirometry in particular, can generate a lot of data. Deciding which data to use for which purpose and how to present the summarized data can be somewhat tricky. This chapter makes some suggestions that may help you publish your results in ways that make sense to the maximum number of people and that may help you avoid committing some common blunders. I am aiming this discussion mostly at newcomers to respirometry. The suggestions below are presented in no particular order. Many are relevant to biomedical data analysis (e.g., metabolic phenotyping; see Chapter 13 for metabolic phenotyping strategies with a more methodological slant).

DO SPECIFY LIVE BODY MASS

Always report the live body mass of your organisms. A surprising number of investigations report only dry body mass, or lipid-free dried body mass, or lipid-free, nonskeletal body mass, and so on, or do not report body masses at all. Your experimental organism probably tightly controls its hydration and lipid levels, and so its total mass—the sum of water, fat, and everything else—*must* be reported. This is not optional. If you have a good reason to believe that much of that mass is water, fat (as in an ob/ob mouse), or food (as, for example, in a late instar caterpillar), also use whatever other unit (such as dry mass) you wish, but do not fail to report the body mass of the intact, functioning organism. Do not fall into the common trap of assuming that fat is an inert tissue with negligible metabolic activity; this widely held but fallacious superstition has been conclusively discredited (Kaiyala et al., 2010).

DON'T USE MASS-SPECIFIC METABOLIC DATA

When reporting metabolic data in any units—milliliters O_2 per unit time, joules per day, microwatts—do not report them only per-unit body mass. If the organisms have a fairly tight body mass distribution (which is to say, if they vary in body mass by only a small amount, and therefore there is no statistically significant relation between body mass and

Measuring Metabolic Rates: A Manual for Scientists. Second Edition. John R. B. Lighton, Oxford University Press (2019).
© John R. B. Lighton (2019). DOI: 10.1093/oso/9780198830399.001.0001

metabolic rate), you can probably divide metabolic data by body mass and get away with it, but you should give the whole-organism figures as well.

If you do express your data in mass-specific form, don't compare the result to another organism of a different mass even of the same species, or to another ontogenetic stage of the same organism. Why? Dividing by body mass makes the implicit assumption that mass-specific metabolic data are not affected by body mass. In other words, you're assuming that metabolic rate scales linearly, that is, isometrically, with body mass. It doesn't; it scales nonlinearly and allometrically with body mass, so mass-specific metabolic data are themselves functions of body mass. See any good intermediate-level physiology textbook for details.

We will be covering this topic in more detail below. It is interesting if depressing to note that the inadequacy of normalizing physiological rate variables by dividing by body mass has been in disrepute for well over half a century (Tanner, 1949), and yet it remains the all but universal practice in biomedical science. At some point in the future a massive tome enumerating all the interpretational fallacies stemming from this approach will be written, but not by me; I try hard to accentuate the positive.

DON'T USE "MASS-INDEPENDENT METABOLIC RATE"

Do not raise body mass to some literature-derived mass scaling exponent, then divide EE by that figure and call it mass-independent metabolic rate. I once received a call from a panicked biomedical researcher who had published in a very prestigious journal (which, interestingly, has accrued no small degree of notoriety among scientists in the know for rushing stories to print that would have benefited from more expert, indeed even marginally competent reviewing) an account that critically depended on comparing metabolism between animals of the same species, but of different masses. The software that analyzed the animals' metabolic data had automatically divided their metabolic rates by $M^{0.75}$. The researcher had only just found this out, and it affected the paper's conclusions.

The value 0.75 is known as "Kleiber's exponent" and is a well-known exponent that (very approximately) relates metabolic rates to mammalian body masses across many species of mammals from shrews to elephants. In other words, it's an *interspecific* mass scaling exponent. The relation between metabolic rate and body mass is completely different *intraspecifically*—within a species (Heusner, 1991). The result, for the above researcher, was a significant distortion of the metabolic data that may have led to incorrect conclusions. Was the study withdrawn? No. Again, metabolic data should only be divided by mass if the mass scaling exponent does not differ significantly from 1.0, or if you don't intend to engage in comparisons. Dividing by mass raised to a mass scaling exponent is a dubious exercise even if the mass scaling exponent is known and not simply plucked out of thin air. If the motive is to compare metabolic data of different organisms of different body masses, a far more powerful technique exists—namely, ANCOVA. See the section "Use analysis of covariance."

DO USE BODY MASS—METABOLIC RATE ALLOMETRY

If possible, report the body mass allometry of your metabolic data. If you are reporting on an organism that varies over a wide range of body masses (say, 25 percent or more), then report on the scaling relationship between the organism's live body mass and its metabolic rate. This is quite simple to do, yet surprisingly few people bother. You merely need to take the logarithms of body mass and of metabolic rate in your preferred units and regress them against each other (Figure 15.1). The slope of the regression is the *mass scaling exponent*, or the power to which body mass must be raised to yield a figure proportional to metabolic rate. For example, a typical mass scaling exponent might be 0.67. A mass scaling exponent of 1.0 means that dividing metabolic rate by mass makes the resulting mass-specific rate independent of body mass. That's unlikely (though certainly possible), and it constitutes the sole circumstance under which dividing metabolic data by body mass is justifiable, aside from the case in which the body mass distribution is too tight to yield a workable data set for allometric analysis and no comparisons involving different body masses are contemplated. Note that the choice of logarithm base does not affect the mass scaling exponent.

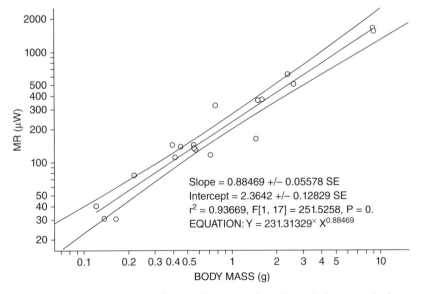

Figure 15.1 An allometric regression showing the mass scaling of metabolic rate on body mass for various species of scorpions from Zzyzx, California. The curves enclose the 95 percent confidence interval of the regression line (ExpeData plot). Raw data are from Lighton et al. (2001).

The intercept of the linear regression relating log-transformed body mass and meta-bolic rate is called the *mass scaling coefficient* after it has been back-transformed by raising the logarithm base to its power. For example, if you are using a base 10 logarithm, then an intercept of 2.00 becomes a mass scaling exponent of $10^{2.00}$ or 100. Thus, for example, the slope of the relation between scorpion body mass and metabolic rate in Figure 15.1 is 0.884, while its intercept (using base 10 log) is 2.364. Ten raised to the power 2.364 is 231, which is the mass scaling coefficient. Thus the allometric equation relating body mass, M, in grams of scorpions (Figure 15.1) to metabolic rate, MR, in microwatts at 25°C is:

$$MR = 231 \times M^{0.884} \tag{15.1}$$

The coefficient of determination for this regression was 0.937, which means that nearly 94 percent of the variation in metabolic rate is explained by variation in body mass.

Always report this relation if your organism's body mass range is wide enough to give you a significant regression. If the body mass range is wide (greater than, say, 50 percent) and the regression isn't significant, then some of the organisms in your sample are prob-ably in different physiological states than others: active versus nonactive, lactating versus nonlactating, male versus female, sick versus healthy, and so on. You will need to reana-lyze your data with this in mind. This level of information is another powerful argument for always investigating the body mass scaling of metabolic rate in your study organisms.

Beware of the possibility that cages with aggressive fan-based convective mixing may cause excessive convective heat loss that is proportional to the surface area of the animal, which scales to $M^{0.67}$. Whether this is a legitimate dependence of EE on M in still air is very debatable—see Chapter 13 for a discussion.

Incidentally, Kissinger's well-known aphorism about academic feuds (they're so intense because there's so little at stake) applies especially to the field of allometric studies of metabolic data. The would-be profundity, blathering, and sectarian warfare prevailing in the field is best avoided. My advice to embrace allometry should not be interpreted as an encouragement to participate in its internecine feuds.

DO USE ANALYSIS OF COVARIANCE

Compare metabolic rates using allometric ANCOVA when possible. Explaining ANCOVA in detail is beyond the scope of this book, but in basic terms it allows you to compare two or more regressions. If those regressions are of log body size versus log metabolic rate, then ANCOVA allows you to compare metabolic parameters between two or more different groups of organisms and determine (1) whether they share a mass scaling expo-nent, and (2) if so, whether their mass-scaling coefficients differ significantly. It does not require you to make any dubious assumptions, and if comparing different groups is not statistically defensible, it will tell you so.

Shortly after the first edition of this book was published, I became aware of an excel-lent treatment of this subject (Arch et al., 2006) to which it is my pleasure to refer you for more details. Also of interest are Kaiyala et al. (2010) and its associated trail of comments in *Diabetes*; Butler and Kozak (2010), Choi et al. (2011), Tschöp et al. (2012), Speakman (2013), and Meyer et al. (2015).

Most of the better statistical packages have an ANCOVA facility, as do some data acquisition and analysis packages. There are also online resources that accept data files from all the commonly used metabolic phenotyping systems: https://calrapp.org/ and https://www.mmpc.org/shared/regression.aspx. Depending on the implementation, the different columns of log-transformed body mass and metabolic rate data either are separated by groups to be compared or are in two long columns separated by a third column that contains a categorical integer (typically) that separates the different groups to be compared. When the ANCOVA is run, it first tests whether the data sets have a common slope (in this case, a common mass scaling exponent). It does so by calculating an *F*-statistic based on the null hypothesis that the slopes are equivalent—in other words, that the data sets can be said to share a common slope. If the probability that the null hypothesis is true, based on the value of the *F*-statistic and its degrees of freedom, is less than (typically) 0.05, or 1 chance in 20, then it can be considered disproved. In such a case, the two data sets do not share a common slope, and that's the end of the comparison. Why? Because if the data sets do not share a common slope, then it is impossible to compare their intercepts (mass scaling coefficients). The data sets are so fundamentally different that no further comparison may be meaningful.

Take that with a grain of salt, though, as the Johnson–Neyman technique (Johnson and Neyman, 1936; see also the excellent review by White, 2003) may offer a solution if the body mass ranges show substantial overlap, in which case it can identify mass ranges that show significant differences between groups.

If, however, the data sets share a common slope, ANCOVA then tests whether the data sets have different intercepts (mass scaling coefficients). It does so by calculating an *F*-statistic based on the null hypothesis that the intercepts are all the same. If the probability that the null hypothesis is true, based on the value of the *F*-statistic and its degrees of freedom, is less than (typically) 0.05, or 1 chance in 20, then it can be considered disproved. In such a case, the two data sets do *not* share a common intercept.

If the data sets share a common intercept, then ANCOVA calculates its value. In such a case, you are justified in pooling all the data, because you have just proved that there's no significant difference between the data sets. If the data sets do *not* share a common intercept, then ANCOVA will tell you the values of the individual intercepts.

Consider the scorpion, a sit-and-wait predator. Is its metabolic rate lower than that of other arthropods? We could take the flat-footed approach and divide everything by $M^{0.75}$, then do a Student's *t*-test, or a Mann–Whitney U test if we are parametric-shy. Fortunately, we know better. We saw in Figure 15.1 that a good allometry between metabolic rate and body mass exists in scorpions. Let's use ANCOVA to compare scorpions against conventional insects and spiders (Figure 15.2). Testing the slopes for homogeneity, ANCOVA tells us that the *F*-statistic is 0.293 at (1, 141) degrees of freedom. There's no significant difference between the slopes; the probability that the slopes are equal is about 50 percent.

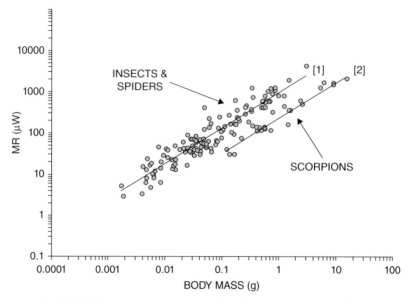

Figure 15.2 An ANCOVA comparing the mass scaling of metabolic rate on body mass for insects and spiders (which share the same scaling; line 1, top) and scorpions (line 2, bottom). Raw data are from Lighton et al. (2001) (ExpeData plot).

The common slope is 0.862. Moving to the intercepts and testing for homogeneity of intercepts, ANCOVA informs us that the F-statistic is 119.3 at (1, 142) degrees of freedom, corresponding to a probability of well under one in a million. The two intercepts are totally different, and we've proved it. But *how* different are the intercepts (i.e., the mass scaling coefficients)? The two coefficients (after transforming them from their logged state) are 977.2 for insects and spiders and 235.5 for scorpions. We have just proved that the mean metabolic rate for a scorpion is about 25 percent that of an equivalent-sized insect, leading to a host of interesting side-effects including cannibalism (see Lighton et al., 2001).

ANCOVA may have handled the case of the heavy and light animals alluded to above with great elegance. Then, the actual results would be in the literature but, of course, might not have been considered novel enough to publish in such a prestigious journal.

DO COMPARE LIKE WITH LIKE

Comparing species that differ widely in evolutionary history, such as the ever-popular rat, chicken, and dog, should be done with great caution. When making such comparisons, you are attempting to draw conclusions from isolated physiological aspects of animals that have progressed along different evolutionary trajectories. Comparing them on, for example, the basis of mass, as if body mass was the only variable differing among them, is ridiculous. To be strictly phylogenetically correct, you need to know and account for the phylogenetic relationship between the creatures you are comparing (Garland et al., 1993; see also Symonds and Elgar, 2002, and references therein). Dragging phylogeny into regression analysis is a deeply complex subject that makes my brain hurt, so I'm not treating it further here.

The alert reader will have noticed that the comparison between scorpions and insects demonstrated above is not phylogenetically correct. This is true. However, enough detailed phylogenetic data to allow a phylogenetically correct comparison did not exist at the time, and the difference between scorpions and other arthropods is so huge that it is impossible for phylogenetic fine-tuning to erase it. As Walt Whitman said, "Do I contradict myself? Very well then, I contradict myself. I am large, I contain multitudes" (Whitman, 1855).

DO LEAVE AN AUDIT TRAIL

Your experimental work is important. It adds to the sum of human knowledge. Do not treat any experimental system as a black box or *deus ex machina* that delivers results you can accept without question. For every item of data you place in a published paper, be able to find in your records the raw data from which it was calculated. Document every step required to reach your final results from the raw data. Store everything that might conceivably be required to link the raw data to your final, processed results. You must be able to locate the raw data that gave rise to every datum you publish and be able to repeat the steps required to reach your published results. Nowadays, data storage space can be regarded as infinite, so use it that way. If you rely on figures spat out by a system that doesn't save the raw data from which the figures were calculated, you are not asking for trouble, you are demanding it, and you deserve your fate.

DO SPECIFY THE TEMPERATURE

Always specify the temperature at which your work was carried out. Ideally, if logistics make it possible, it is best to control the temperature of your organism's environment. If you can't control it, at least report it. And when you report it, let it be generated from something other than a supermarket toy made by a political prisoner on the short list for organ donation. Take temperature (whether controlled or ambient) seriously and record it as one of your experimental variables with your data acquisition system. Report real means and standard deviations, not just guesses. Ideally, use a National Institute of Standards and Technology (NIST)-traceable temperature measuring instrument.

As covered in more detail below, temperature profoundly affects the MR of heterothermic ectotherms (organisms that take on the temperature of their surroundings). Those who have not taken a formal course in integrative animal biology might be surprised to learn that the MRs of homeothermic endotherms (organisms that maintain a

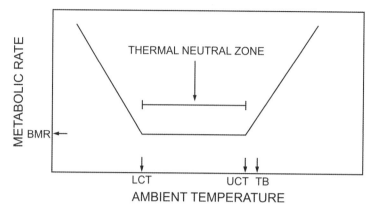

Figure 15.3 The effect of ambient temperature on an endothermic homeotherm, an animal that can generate significant body heat and that regulates its body temperature (TB), such as a typical mammal. Between the lower critical temperature (LCT) and the upper critical temperature (UCT), the animal's MR is at its lowest. Between these limits, which form the boundaries of the thermal neutral zone, the animal can regulate its TB with minimal metabolic cost by changing its posture, peripheral vasomotor tone, and hair or feather orientation. If the animal is postabsorptive (fasting), that low MR is its basal metabolic rate (BMR). Below LCT the MR increases because of shivering, BAT induction or other forms of thermogenesis; above UCT (which is always less than the animal's regulated TB), the MR increases because of sweating or other heat-loss strategies.

fixed body temperature higher than their surroundings, such as mice and humans) are also affected by temperature, as Figure 15.3 makes clear. Thus, measurement temperature must *always* be stated. If this is news to you, consult an intermediate-level textbook of animal physiology, and the brief description that follows, for more details. A good representative review of temperature effects on small-mammal metabolism can be found in Rezende et al. (2004).

Consulting Figure 15.3, we see that the MR of a homeothermic, endothermic animal such as a mammal or bird will reach a stable minimum (basal metabolic rate (BMR)) over a range of temperatures from the lower critical temperature (LCT) to the upper critical temperature (UCT). This follows directly from the fact that the animal is attempting to defend a stable body temperature, typically 37°C. At cold temperatures, excess metabolic energy must be generated in order to maintain that body temperature in spite of an elevated thermal gradient driving heat out of the animal. As ambient temperatures increase, the gradient declines until passive mechanisms can conserve body heat without requiring additional energy expenditure. This marks the beginning of the LCT in which three primary mechanisms conspire to retain body heat. These are postural adjustments (increasing surface area to lose heat, or reducing it to conserve heat), increasing or reducing insulation (by puffing out or flattening fur or feathers), and increasing or decreasing vasomotor tone (transporting heat to the skin surface by peripheral vasodilation, or keeping it near the core of the body by peripheral vasoconstriction). Finally, as ambient temperature rises to a value slightly below the defended body temperature, these predominately passive mechanisms fail, and more energetically expensive tactics are required in order to lose additional body heat. This is accomplished primarily by evaporation of water, often accompanied by muscular activity designed to increase convection and enhance evaporation by disrupting boundary layers.

Generally speaking, the LCT of a mammal or bird will increase as its body mass decreases, because it is generating body heat with a smaller mass of tissue while its surface area to volume ratio, which determines its rate of heat loss, increases. For example, a standard laboratory mouse strain such as the C57BL/6J shows an LCT of about 29°C (Figure 15.4). The alert reader with a biomedical background will recognize that such animals, which are among the most important model animals in the field, are usually investigated at a so-called room temperature of about 21°C (if the ambient temperature at which the investigation took place is reported at all, which is all too seldom the case). At such temperatures the animals are severely cold-stressed. They are generating a large amount of metabolic heat simply in order to defend their core body temperature. Their subjective impression of comfort, as determined by voluntary preferences in a temperature gradient, would be far greater at a temperature closer to their thermal neutral zone (TNZ). It is amusing, or perhaps depressing, to note that IACUCs regard 29°C as a savagely cruel ordeal for the poor mice, requiring extraordinary justification for its approval in an experimental protocol. Yet this is the temperature that the mice will voluntarily choose if they are given the choice (Gordon et al., 1998). Talk about macromammalian chauvinism!

A note on temperature measurement: This is nontrivial, because the "real" or operative temperature (Bakken et al., 1985) to which an animal is exposed consists of a mixture of conductive, convective, and radiative components. Fortunately, in a laboratory environment a simple temperature probe in or near the cage(s) will usually suffice. In more challenging environments with substantial convective and radiative components, it may be necessary to construct a model of the animal (ibid.) and measure the model's temperature.

To a good approximation, the data in Figure 15.4 allow us to predict, approximately, the mean daily energy expenditure (DEE) of a typical 30-g mouse across the normally encountered range of ambient temperatures below thermoneutrality, especially in the range of 20–30°C. Thus, where ambient temperature in degrees Centigrade is denoted by T,

$$\text{kcal h}^{-1} = 0.959 - 0.0225 \times T \tag{15.2}$$

$$\text{Watts} = 1.113 - 0.0261 \times T \tag{15.3}$$

$$\text{mL O}^2 \text{ min}^{-1} \sim= 3.30 - 0.0774 \times T \tag{15.4}$$

The latter equation assumes an RQ of 0.83, yielding an oxyjoule equivalent of 20.23 mL J O_2^{-1} from equation 9.13. The above three equations should provide easy translation into any commonly used units.

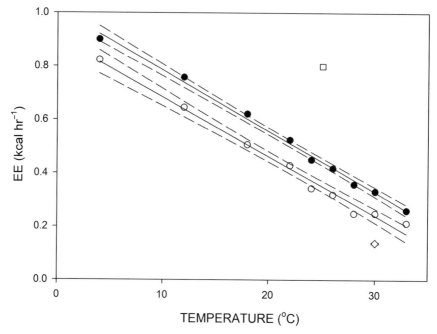

Figure 15.4 The relation between ambient temperature and daily energy expenditure (EE) in chow-fed C57BL/6J mice, body mass approximately 30 g (nocturnal phase = closed circles; diurnal phase = open circles). The 95 percent confidence bands of the regressions (dashed lines) are shown. Data were digitized from Figure 3C of Abreu-Vieira et al. (2015). The huge effect of ambient temperature is clearly seen. See text for predictive equations. Outliers: Data are from an elderly metabolic phenotyping system with a faulty flow rate meter (square; Ogimoto, personal communication) and indirect calorimetry data from a comparison of direct versus indirect calorimetry (diamond; digitized from Burnett and Grobe, 2013, Figure 3C; 2014, Figure 1B; standard error bars are within the symbol).

For diurnal data only, which consist mostly of resting energy expenditure (REE), eq. 15.2 becomes

$$\text{kcal h}^{-1} = 0.905 - 0.0222 \times \text{T}$$

$$(15.5)$$

while for nocturnal data only, with a much larger active energy expenditure (AEE) component, eq. 15.2 becomes

$$\text{kcal h}^{-1} = 1.013 - 0.0228 \times \text{T}$$

$$(15.6)$$

If you are measuring the metabolic rate of an orthodox C57BL/6J mouse and your data are more than ~ 25 percent away from the predicted value at an ambient temperature between 20 and 30°C, you may wish to pause for reflection and investigate further. Two examples are shown in Figure 15.4. The elderly metabolic phenotyping system (empty square) shows an ~ 45 percent error in daily EE. The indirect calorimetry data from Burnett and Grobe (2013 and 2014; empty diamond), which they report as underestimating their direct calorimetry data, is ~ 60 percent of the predicted value based on minimally active (diurnal phase) mice at the measurement temperature.

As a further complication, the ingestive history of the animal affects its EE. Following ingestion (particularly of protein), a transient elevation of EE occurs, variously referred to as specific dynamic action (SDA) or diet-induced thermogenesis (DIT). It is for this reason that the BMR of an animal is measured in its post-absorptive state.

Into this complex world of temperature and other effects wades the unwary researcher who wishes to compare the metabolic rates of two groups of animals, one a wild type and the other a gene knock-out. A well-known side-effect of genetic manipulation is reduction in body size; another is thin, patchy, or even absent fur (B. Cannon, personal communication). Our researcher, blithely unaware of the physiology of thermoregulation, now compares these two groups of animals at 21°C, a temperature well below the animals' LCT—meaning that they are cold-stressed. Behold, the knock-out mouse is found to have a significantly higher metabolic rate. This is not surprising, because it is smaller and thus has a larger surface area (which loses heat) to volume (which generates heat) ratio. Therefore, it requires a higher metabolic rate (1) in order to defend its body temperature, which we assume to be the same as the control group; and also (2) because it is losing heat more rapidly through its thin and patchy fur, and thus requires a concomitant increase in metabolic rate to compensate for this extra avenue of heat loss. Thus, the published results of the study say nothing about intrinsic differences in metabolic rates. They are byproducts of the physics of biological heat transfer in two groups of animals with significant though unrecognized differences in thermal physiology, and measured under circumstances that make metabolic comparisons almost meaningless.

What to do? There is one very simple solution, which is to determine the LCTs of the two groups and compare them within each group's known TNZ range, and, if possible, at the same temperature (see Meyer et al., 2015). In the absence of detailed knowledge, using a consensus temperature of, say, 31°C would be reasonable for a metabolic comparison study of mice or rats if your IACUC will allow it. This allows a direct and rigorous comparison of the two groups, especially if ANCOVA is used to remove the effect of different body mass ranges.

DO STATE THE TEMPERATURE EFFECT
ON ECTOTHERMS CORRECTLY

By all means necessary, analyze temperature effects properly. Most readers of this book suffered through, or even enjoyed, a course or three in physical chemistry. There is absolutely no reason for the almost universal practice of using a linear regression to express the effect of temperature on metabolic rate. You know that temperature has an exponential effect on the rate of any single enzyme system, including those involved in metabolism, at least up to the point of enzyme denaturation or O_2 limitation. If your

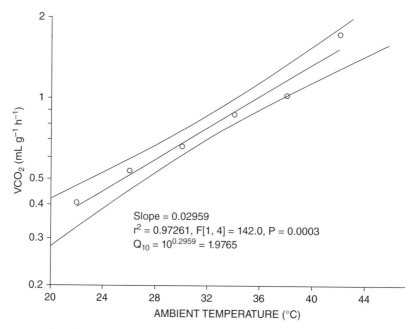

Figure 15.5 The effect of temperature on the metabolic rate of a harvester ant. The curves enclose the 95 percent confidence interval of the regression line (ExpeData plot). The slope corresponds to a Q_{10} of 1.97, close to the commonly assumed value of 2.0. Raw data are from Lighton and Duncan (2002).

organism displays an apparently linear effect of temperature on metabolic rate, this simply means that natural selection has favored your organism's ability to modulate its temperature sensitivity as a function of temperature. When you do graphic analyses of ectotherm metabolic rate versus temperature data, use a linear X axis for temperature and a log-transformed Y axis for metabolic rate (Figure 15.5). If the result is a more-or-less straight line, your organism has a consistent temperature sensitivity of metabolism. You can calculate the temperature sensitivity from the slope (see the next section). If the slope is curvilinear, you can fit a polynomial to it and calculate temperature sensitivity as a function of temperature from its first derivative (see the next section).

DO USE A SENSIBLE TEMPERATURE SENSITIVITY METRIC

The most intuitive measure of temperature sensitivity is the Q_{10}, or the ratio of metabolic rate at a given temperature divided by the metabolic rate at a temperature 10°C lower. It is relevant mainly to ectothermic and heterothermic organisms—that is, organisms that do not keep their body temperature constant by modulating their metabolic rates, and whose body temperature is consequently variable, such as most arthropods, amphibians, and reptiles. A Q_{10} of 2 means that an organism doubles its metabolic rate with every 10°C increase in temperature. That figure gives you an intuitive feel for the organism's temperature sensitivity.

Strange informational dynamics exist in science. One good example is that facts in organismic biology, such as the temperature sensitivity of organisms, not to mention the fact that an organism's metabolic rate depends on its size, will drift into other fields of science such as ecology, where they will be buried and resolutely ignored (The mere *plumbing* of an organism! How *irrelevant*, not to mention how *gross!*) for decades. Finally, these facts reemerge as will-o'-the-wisps that the natives of the field believe to be local phenomena. They proceed to invest them with uncanny and irrational meaning, even worshiping them as new-found gods in a bizarre inversion of the cargo cult. One such will-o'-the-wisp lately drifting back into organismic biology from ecology favors expressing the temperature sensitivity of metabolism using electron volts of activation energy. This is ridiculous for three reasons. First, it is pretentious physics envy. Second, it is not intuitive. Third and most important, it is ignorant. Overall, metabolism is the summed effect of many enzymes and pathways interlocked in exquisite tangles that cannot be legitimately summed up as a single activation energy. To posit one activation energy for the summation of the process displays breathtaking ignorance. Please respect the complexity of your study organism and the desire of those who read your work to understand the temperature sensitivity of your organism using an intuitive, empirical metric such as Q_{10}. Leave electron volts (and the word *quantum*, please) to people who understand how and where to use the term.

An organism's Q_{10} is easily calculated by plotting log-transformed metabolic rate against temperature. The slope of the line is multiplied by 10, and the logarithm base is raised to that power. As an example, consider a regression of log metabolic rate in any units against temperature in degrees Celsius. If the slope is 0.0312 and the logarithm base is 10, then for every increase in temperature of 1°C the creature increases its metabolic rate by a factor of $10^{0.0312}$, or 1.074. To get the Q_{10}, we increase the span from 1 to 10°C by multiplying the exponent by 10; $10^{0.312}$ equals 2.051.

If the relation between log metabolic rate and temperature is curvilinear, note the oversimplification of expressing temperature sensitivity as a constant, let alone the absurdity of then expressing it in electron volts. If you would like to quantify the change in temperature sensitivity as a function of temperature, simply run a polynomial regression on the data, take the first derivative of the polynomial (thus creating the slope of the curve as a function of temperature), multiply that vector by 10, and raise your logarithm base to the power of the transformed vector (Lighton and Bartholomew, 1988).

If you know the Q_{10} of an ectothermic organism, you can take a set of metabolic rates measured at various temperatures and convert them to a single temperature. The formula for doing so is:

$$MR2 = MR1 \times Q_{10}^{[(T2-T1)/10]} \tag{15.7}$$

where MR2 is the metabolic rate standardized to temperature *T2* (in Celsius or Kelvin), 2.0 is assumed for Q_{10} if it is not known, MR1 is the measured metabolic rate, and *T1* is the temperature at which MR1 was measured.

DO SELECT METABOLIC DATA THOUGHTFULLY

Especially in flow-through respirometry, which gives you a vivid picture of an animal's metabolic state as a function of time, you may be faced with an abundance of data and need to employ replicable, objective selection criteria when deciding how to choose sections of recordings for later analysis. Much talk centers on selecting a low, stable reading of a certain length, to ascertain REE. Hayes and colleagues (1992), in their excellent treatment of this topic, state that "what constitutes a 'low' or 'stable' value is subjective and potentially a source of substantial bias. The longer an animal's metabolism is monitored, the lower the animal's metabolism . . . is likely to be" (p. 605).

Ultimately, we are usually after two things. We want a low value of metabolism that reflects metabolic processes without the complication of paying the price for motor effectors to kick in; in other words, the animal should be as motionless as possible. We also want a value of metabolism that is stable; in other words, it should be in equilibrium, neither increasing nor decreasing during the measurement. For example, DIT following food ingestion might interfere with REE measurement, as might fluctuations in temperature (at least outside the range of the TNZ).

Activity detection (Chapter 18; see also Chapter 13) provides an objective assessment of whether activity is taking place and should always be recorded where possible. Recordings that include activity data allow sections of the recording to be identified that should not be analyzed further if low, stable metabolic data are required. (Of course, activity metabolism may be desirable data for other purposes, in which case its objectively verified presence is useful.) Given that the animal was inactive, how do we proceed?

Hayes et al. (1992) scanned recordings for the lowest interval of 15 min during periods ranging from 30 min to 6 h after starting the recording. As the overall duration increased, the lowest section of 15 min decreased in magnitude. The decrease was rapid at first, approaching an asymptote after about 4 h. Thus, when reporting metabolic data of this kind, it is very important to determine what degree of sampling bias you are introducing by determining the interplay of recording duration and search window. Good data analysis software has the capability to define a search window width and find the corresponding section of the recording with lowest (or highest, or most level) readings, so this is not difficult to do in practice. It is important to specify, when publishing, the exact objective criteria that you used.

The recommendation therefore is to monitor individual animals for as long as practical and to find objectively a low, level section of metabolic data for each. Withers (2001) favors examining the distribution of residuals around a linear regression of the recording; I would recommend instead using a sliding window of a defined width and finding the location in the recording where the window has the lowest variance and lowest magnitude.

More recently, Cooper and Withers (2010) have added informatively to the uncertainty in this area by demonstrating that the metabolic sampling causes small but significant effects on minimum metabolic rate measurement.

Cooper and Withers (2010) did so by comparing four different regimens: "continuous" (sampled every 20 sec, which strikes me as somewhat coarse-grained; I would recommend sampling at 1 Hz) and at subsamples picked every 3, 6, and 12 min from the continuous record. This guaranteed that all sampling regimens used equivalent data, while making the implicit assumption that the subsampled data suffered no loss of accuracy in the act of subsampling (for example, by not allowing adequate time for analyzers to stabilize following switching of the gas streams). They then calculated minimal values (as for a basal metabolic rate study) over periods of 21, 36, and 60 min for all regimens.

When averaged over equivalent time periods, all intermittently sampled regimens somewhat underestimated both VO_2 and VCO_2 relative to the continuous sampling regimen. However, as Hayes et al. (1992) point out, prolonging the duration of minimal values causes an increase in mean minimal values, so to an extent this effect can be canceled by averaging over longer intervals. The accuracy penalty is also quite small (almost always < 10 percent and usually < 5 percent). Nevertheless in their hands, and with their sampling intervals, continuous recordings provide better data than do intermittently sampled recordings. However, continuous recording can be prohibitively time-intensive if only a single animal can be measured at a time, as is the case in many real-world continuous monitoring systems, tempting investigators to record metabolic data over shorter periods than are required for an animal to acclimate fully to the respirometry chamber. And although multichannel continuous metabolic phenotyping systems are commercially available, they are significantly more expensive than multiplexed systems and require carefully matched calibrations between a teeming swarm of gas analyzers, somewhat limiting their use. Cooper and Withers (2010) conclude that much of the metabolic data for small mammals may "be overestimated by 15%–28%…because of insufficient experimental duration" and that "compared with these sources of error, the error from an interrupted sampling regime appears to be low to intermediate." Of course, other, unmodeled effects may worsen this error.

For a more biomedically oriented treatment of REE, DEE, and other topics, the reader will find Meyer et al. (2015) to be an excellent summary. It is worth mentioning that the majority of metabolic phenotyping systems used in biomedical research have very long cage time constants ($\sim$ 20 min; see Chapters 8 and 13). This complicates determination of REE, and especially compromises accurate measurement of AEE. EEs corresponding to periods of activity and inactivity tend to be blurred together (see Figure 15.4, bottom line). Some rather complex algorithms have been derived to cope with this instrumental shortcoming (e.g., Van Klinken et al., 2012), but it is better, in my opinion, to solve this problem at the root by using a system with the shortest possible time constant in the first place. Meyer et al. (2015) do a good job of pointing out some interesting subtleties in REE measurement, such as the anomalous dip in EE that often follows bouts of activity and is caused by the temporary relaxation of metabolic thermogenesis, in turn caused by a fall in core body temperature following activity (see also Meyer et al., 2004; Speakman, 2013).

A more methodologically oriented treatment of this subject can be found in Chapter 13.

DO SANITY CHECKS

Congratulations! You have completed a round of metabolic measurements with your experimental animals. You send in the paper and await feedback with eager anticipation. Yet somehow an experimental error crept into your measurements, and the reviewers politely (or not so politely!) point out that your control-group metabolic measurements are substantially at odds with values previously reported by respected investigators. A sanity check on your results might have saved you a great deal of trouble, especially if the reviewers were less than diligent and the paper was accepted for publication.

As an example, imagine that you are working with mice in a biomedical investigation. Are your EE measurements in line with predicted results? Figure 15.4 shows you one method for determining that. A literature search will almost certainly yield useful results for the animal taxon you are investigating (for example, see Lighton et al., 2001, for the EE of insects, ticks, and scorpions as a function of body mass at 25°C). If your data are substantially at odds with the consensus in the field, this does not necessarily mean that they are incorrect. It does, however, mean that you should take extra care to ensure that your measurements are accurate and defensible. It is a wise choice to buttress your arguments and findings with proactive validations before being required to do so by a journal editor or—worse—by researchers attempting to replicate your work after publication.

16

The Varieties of Gas Analyzers

In this chapter I describe the most common gas analyzers for determining the partial pressure of the most common gas species in respirometry: O_2, CO_2, and water vapor. Aquatic gas analysis is treated separately in Chapter 5. Each kind of analyzer has its own characteristics, strengths, and weaknesses; knowing them will help you get the best results from your respirometry system. Knowing their characteristics is especially important if you intend to use flow-through respirometry.

All O_2 and CO_2 analyzers actually measure the partial pressure of O_2 or CO_2 and span this figure (i.e., multiply it by a constant that the user interactively sets during calibration) to read a concentration, typically in percent. Now, if the local barometric pressure changes, so will the po_2 and pco_2. The analyzer's readings will reflect this, even though the concentration of O_2 or CO_2 has not changed. Good-quality analyzers correct their concentration measurements for changes in barometric pressure, which is a major advantage in research applications.

If you are going to purchase an analyzer, consider when making your decision not only the kind of analyzer, but also some additional factors such as: Does the manufacturer have a track record supplying analyzers for research (rather than instructional) respirometry? Can it provide in-depth, research-grade technical support for the use of its analyzers in this area? Does it have "street credibility" or does it betray low ethical standards by making exaggerated claims or by emphasizing its low cost while claiming high quality? What is the length of its guarantee? Don't be shy about asking around or about asking a manufacturer for citations or references to satisfied customers in your field. Search on Google Scholar for citations of the company's equipment. Always remember that the price of research-grade equipment is negligible against the context of a research program that must produce good results and bring in multiple grants across many years. Penny-pinching on equipment may end up being an expensive, embarrassing, and professionally damaging decision.

OXYGEN ANALYZERS

Because respirometry measures tiny deviations from a massive atmospheric O_2 signal, it is an extremely demanding application for any O_2 analyzer. Choose your analyzer with

Measuring Metabolic Rates: A Manual for Scientists. Second Edition. John R. B. Lighton, Oxford University Press (2019).
© John R. B. Lighton (2019). DOI: 10.1093/oso/9780198830399.001.0001

care. There are three main varieties of aerial O_2 analyzers; paramagnetic, zirconia cell, and fuel cell.

Which O_2 analyzer to use depends on your budget, your application, and your level of patience. I discuss the three major types in turn. Incidentally, fluorescence-quenching-based and tunable-laser-based aerial O_2 analyzers do exist, but in my opinion they do not have adequate resolution for research-grade flow-through respirometry, and so will not be considered further.

Paramagnetic oxygen analyzers

Paramagnetic O_2 analyzers make use of the paramagnetic properties of O_2, discovered by Michael Faraday at the Royal Institution in London, who at one point was the laboratory assistant of my maternal great-great-grandfather, William Thomas Brande, who is now almost entirely forgotten. If a sealed vial filled with O_2 is suspended from a quartz thread, it will be attracted to a strong applied magnetic field. Note that bulk O_2 itself does not have a net magnetic field; being paramagnetic, rather than magnetic, it requires an external magnetic field to align its constituent magnetic dipoles before its magnetic properties can manifest themselves. Nitrogen, in contrast, is opposite; it, too, requires an external magnetic field to manifest its magnetic properties, but it is repelled from magnetic fields and is referred to as diamagnetic.

The paramagnetic O_2 analyzer in its usual configuration does not make use of O_2-filled but rather N_2-filled vials to maximize their interaction with O_2. The vials are usually tiny spheres, about 2–3 mm across, mounted like the weights of a dumbbell on each end of a thin rod. The rod is suspended in a strong, uneven magnetic field by a quartz thread, and at the rod's center, where it meets the thread, it holds a tiny mirror. The geometry of the chamber holding this assembly is arranged so that incoming air will buffet the delicate dumbbell as little as possible.

As incoming O_2 levels in a gas sample rise, the dumbbell generates an increasing torque because the gas sample surrounding it becomes more and more paramagnetic on one side (remember that the magnetic field is uneven). Early paramagnetic O_2 analyzers simply allowed the dumbbell to rotate and used the mirror to move a spot of light on a ground-glass scale, graduated in percent O_2. This was suboptimal because as the dumbbell moved in the magnetic field, its magnetic environment changed and its response to O_2 became nonlinear. Modern paramagnetic analyzers incorporate a small coil in the dumbbell assembly, transforming it into a tiny and very weak electric motor. A light source and photodetector assembly detect any deviations of the dumbbell assembly from its null position and adjust the current running through the coil to rotate the dumbbell back into alignment. The current that accomplishes this task is very precisely a function of the paramagnetic content of the gas sample. With care, accuracy levels of 0.02 percent can be achieved (Kovacich et al., 2006).

This works well if O_2 is the only paramagnetic gas in the sample. Fortunately, very few gases are paramagnetic enough to interfere, except for nitrous oxide, which is almost half as paramagnetic as O_2. Also, water vapor is diamagnetic and will therefore interfere with O_2 measurement. Therefore, paramagnetic analyzers are not recommended for use with mathematical water vapor dilution compensation. For general nonanesthesiology respirometry in a stable laboratory environment, the paramagnetic analyzer works outstandingly well. One particularly welcome aspect is that its lifetime is indefinite if it is treated with respect. However, the paramagnetic analyzer has a few faults, and these can be significant in a typical working metabolic measurement laboratory.

It is fragile and very sensitive to mechanical shocks and vibration. Even if they are not strong enough to break it, such stresses will cause reading jitter. Moving things around on the same bench top will have noticeable effects. You will have to learn to walk on eggs a bit.

It is also very sensitive to flow rate variations. When you calibrate a paramagnetic analyzer, you *must* zero and span it (see the section "Calibrating a paramagnetic oxygen analyzer") and read your sample at the exact same flow rate. The flow must be stable.

It can be damaged by too high a sample flow rate, for obvious reasons. Typically, sample flow rates are limited to about 200 mL min^{-1} at most. This, combined with its sensitivity to flow-rate variations, means that paramagnetic O_2 analyzers are almost always used in subsampling mode with a dedicated subsampling pump and flow control system, either internal or (preferably) external.

It is sensitive to temperature fluctuations because its magnets are affected by temperature. All paramagnetic analyzers have temperature compensation circuitry that reduces this effect, but it can still be significant. Some large industrial analyzers use internal ovens to maintain their paramagnetic bench at a fixed temperature, but in typical laboratory settings this is seldom required.

Calibrating a paramagnetic oxygen analyzer

To calibrate a paramagnetic analyzer, you need two gases: one is pure O_2 or (preferably) outside air scrubbed of water vapor and CO_2, and the other contains no O_2 at all (typically nitrogen). You should be aware that nominally pure or "medical grade" O_2 may, in fact, be anywhere from 93 to 99.5 percent pure; this is fine if you know the actual concentration from the manufacturer, failing which you will need to specify (and pay for) 100 percent O_2. Outside air scrubbed of water vapor, with a nominal O_2 concentration of 20.94 percent, is more convenient and much cheaper (see the section "Calibrating a fuel-cell oxygen analyzer").

Pull the sample through the analyzer at about 150 mL min^{-1} using a subsampling pump and flow meter. This flow rate does not have to be accurately known, but it must be reasonably stable. The input of the analyzer is connected to a manifold, which may be as simple as an empty syringe barrel (about 10 mL capacity; see Figure 16.1). A sample flowing into the barrel can be subsampled without changing the analyzer's flow rate or internal pressure. First bleed nitrogen into the manifold at a flow rate of about 300 mL min^{-1}; the rule when using manifolds is to allow air to enter them with at least twice the flow rate than air is withdrawn from them. Zero the analyzer (see the manufacturer's

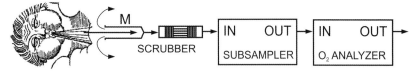

Figure 16.1 Using a manifold to subsample a low flow rate of analyte gas from a higher-flow air stream without pressure effects.

instructions). Now flow dry, CO_2-free outside air (20.94 percent) or another known concentration of O_2 through the analyzer and adjust the analyzer's span reading appropriately. This is referred to as *spanning* the analyzer. Repeat until the analyzer moves gracefully between the zero and span gas concentrations when they are alternated. The sample air stream can now be pumped into the manifold. Note the emphasis on keeping the flow rate through the analyzer constant by using the manifold to isolate it from the rest of the system.

Several variants of paramagnetic gas analyzers exist. Most nondumbbell types are used for industrial, medical, or sports physiology gas analyzers where large signals are the rule, and they tend to be noisier, less accurate, and more drift-prone than the magnetodynamic or dumbbell paramagnetic analyzer described above.

Zirconia-cell oxygen analyzers

Zirconia-cell O_2 analyzers use the principle that, when heated, zirconium oxide ceramic becomes a solid-state electrolyte through which O_2 molecules can migrate. The ceramic is formed into a hollow cup, with one side typically exposed to air and the other to the sample gas. As the O_2 molecules migrate across the ceramic wall of the cup, from the higher to the lower po_2, they create an electric charge across the wall. The charge is picked up by a porous platinum film on either side of the cup. The resulting signal obeys the classic Nernst equation:

$$mV = (RT)\left[\ln\left(p_1O_2 / p_2O_2\right)\right]/(nF) \qquad (16.1)$$

where R is 8.3144 J $(mol\ K)^{-1}$, T is the temperature in Kelvin, n is the number of electrons dragged by each O_2 molecule (four), F is the Faraday constant, p_1 and p_2 are two partial pressures of O_2 on either side of the cup, and mV is the potential difference across the cup wall in millivolts.

This all takes place at about 650°C, which requires a lot of energy to reach and maintain. Thanks in part to the high temperature, O_2 moves at blazing speed through the ceramic, so the response time of zirconia-cell analyzers is extremely fast. In addition, as the O_2 potential pressure difference across the cup wall climbs, the signal from the cup gets larger and larger (rather than smaller and smaller), making the zirconia analyzer effective at measuring very low O_2 concentrations if ambient air is used as the reference gas, which is usually the case.

Fast response times and excellent low-range performance are the zirconia analyzer's two strong points. The latter is unimportant in respirometry as we know it, however, and unless you need to do breath by breath analysis (which is a much-overrated concept; see Chapter 12), so are very fast response times. This is because the limitation on the response times in most respirometry systems is set primarily by the flow rate divided by the volume of the chamber and tubing, which is typically anything from several seconds to a minute or more, or, in the extreme case of room calorimetry (Chapter 12), several hours. In the vast majority of respirometry setups, all three major types of O_2 analyzers will produce results that are difficult or impossible to distinguish on the grounds of response speed, even though the native response speeds of the analyzers themselves may be very different. Where the differences will show up in practice are in the areas of noise, stability, and service life, in all of which zirconia-cell analyzers have shown themselves to be undistinguished.

The primary disadvantages of the zirconia-cell O_2 analyzer are as follows.

It is bulky, requires high power (*not* for field use!), and requires a long warm-up period, typically 4 h or more.

It is more prone to drift and noise than the other two types of analyzers.

It is very sensitive to changes in flow rate.

If liquid water enters its zirconia cell, the water will explosively change phase into steam, and the pressure front may crack the cell.

If volatile organic compounds or gases such as methane or acetone enter the cell, they will combust with atmospheric O_2, causing an artificial increase in O_2 depletion. With some organisms this can result in massive measurement errors, often in the form of subphysiological RQs.

Over time, the sensor will (depending on the manufacturer) degrade and must be "rebuilt" periodically at considerable cost. Zirconia-cell analyzers are often marketed as not having a limited lifespan, but this is far from true; 30,000 h or about 3.4 years of continuous use is typical.

Thus, though an excellent choice for industrial trace oxygen gas analysis, the zirconia-cell analyzer does not have clear advantages in most practical respirometry systems, and it may have substantial disadvantages. Instructions for zeroing and spanning zirconia-cell analyzers differ widely from manufacturer to manufacturer and are not discussed here, other than to mention that because of their high sensitivity to flow rate, these analyzers give their best performance if a subsampling system is used (similar to that described for the paramagnetic O_2 analyzer). Differential zirconia-cell O_2 analyzers which are capable of somewhat better noise and drift performance do exist, but they are very expensive.

Fuel-cell oxygen analyzers

Fuel-cell O_2 analyzers use replaceable electrochemical cells to sense O_2. These are often called *fuel-cell sensors*. O_2 diffuses through the front membrane, typically made of Teflon (see Appendix 2), travels through an electrolyte, and oxidizes a lead anode, so:

$$\text{Cathode reaction: } O_2 + 2H_2O + 4e^- \rightarrow 4OH^- \qquad (16.2)$$

$$\text{Anode reaction: } 2Pb + 4OH^- \rightarrow 2PbO + 2H_2O + 4e^- \qquad (16.3)$$

The overall cell reaction is as follows:

$$O_2 + 2Pb \rightarrow 2PbO \qquad (16.4)$$

Thus, each O_2 molecule liberates four electrons. This tiny current is sensed and amplified in the analyzer's circuitry, and in good-quality analyzers, corrected for changes in barometric pressure. The fuel-cell sensor is temperature sensitive and contains an intrinsic thermistor-based thermal compensation array to make the output of the cell nominally constant over temperature. For critical applications, the fuel cell can be placed in a thermally regulated environment.

One major advantage of the fuel cell is the fact that its resolution can be extraordinarily high and is primarily limited by the skill with which its signal can be amplified and digitized. As an example, a dual-absolute, temperature-controlled O_2 fuel-cell analyzer (Sable Systems Oxzilla) has attained resolution of 0.3 ppm O_2 against a background O_2 concentration of ~ 209,400 ppm (Stephens et al., 2007). This makes the fuel-cell O_2 sensor an ideal choice—if well implemented!—for flow-through respirometry, where high resolution and low noise are critically important.

Because the fuel cell is not heated to a high temperature and does not contain components that can be disturbed by flow, O_2 analyzers based on the fuel-cell principle can be used over a very wide range of flow rates, typically 10–2000 mL min^{-1}. Although readings do change with flow rates because any air flow requires a small pressure differential which translates to a change in the partial pressure of O_2 to which the analyzer responds, this effect is marginal except at very high rates of flow. Nevertheless, because the fuel-cell O_2 analyzer (like all O_2 analyzers) responds to changes in partial pressure rather than absolute concentrations of O_2, it should generally be placed last in an analyzer chain and air should be pushed through it rather than pulled, with the exit port open to the ambient air. This is especially the case if periodic alterations of flow resistance occur in the gas path, such as when switching a solenoid valve from baseline to sample. If air is pulled from an O_2 analyzer through such a valve, baseline offsets are inevitable because the two pathways through the valve will differ in flow resistance. Placing the valve upstream from the analyzer, pulling air from it with a subsampling pump, and then pushing the air into the O_2 analyzer will isolate these resistance changes, which will no longer affect the O_2 readings.

Being robust, unfussy, and reliable, the fuel-cell O_2 analyzer is a good choice for both general and precision respirometry. It does have some disadvantages, and these include the following.

The fuel-cell sensor will last for about 2–3 years before needing to be replaced. If your fuel-cell-based O_2 analyzer is noisy, unstable, and just plain difficult to operate (but was made by a reputable manufacturer), it needs a fuel-cell replacement. Fortunately, replacing the fuel cell is easy and inexpensive with good-quality analyzers. It is interesting to note that zirconia-cell analyzers have a practical lifetime only marginally longer than fuel-cell sensors, requiring a regular factory refurbishing at much greater cost and inconvenience.

The time constant of a typical fuel-cell O_2 analyzer is about 7–10 sec for 10–90 percent of a step change. This is not a limitation in general respirometry. This disadvantage can be overcome in large part by using the "instantaneous correction" described in Chapter 8. Fuel cells with faster response times but shorter lifespans may be available from the manufacturer.

Although the thermal compensation array reduces temperature effects, it does not eliminate them. Because the array and the fuel-cell electrolyte have different thermal time constants, fuel-cell O_2 analyzers are particularly sensitive to the *rate* of temperature change. A sudden change of temperature may create significant drift until the cell's electrolyte and its thermal compensation array re-equilibrate. Good-quality analyzers use proprietary thermal buffering techniques to reduce this effect. If you *must* use your fuel-cell O_2 analyzer in unstable conditions, frequent baselining will practically eliminate thermal drift. Again, for precise applications, thermal regulation of the fuel cell's environment is recommended. Fuel-cell O_2 analyzers that are internally temperature-controlled eliminate most of the above disadvantages and offer outstanding stability, but are quite expensive.

Fuel-cell O_2 analyzers are very slow responding at very low or very high O_2 concentrations, so using them below 1 percent or above 95 percent O_2 is not advisable. Paramagnetic analyzers are a good choice for high-end concentrations, and zirconia-cell or paramagnetic analyzers are good for low-end concentrations.

Calibrating a fuel-cell oxygen analyzer

To calibrate a fuel-cell O_2 analyzer, you need outside air scrubbed of water vapor or, less satisfactorily, 100 percent O_2. (Be aware that "medical grade" O_2 can be anywhere from 93 percent O_2 on up, so do not rely on its purity for calibration purposes.) Push the air through the analyzer, leaving its exit port unconnected so that the sensor is close to ambient barometric pressure, and adjust its span setting until the display matches the span gas concentration. In the case of dry outside air this will be 20.94 percent with a negligible margin of error (Tohjima et al., 2005). This is referred to as *spanning* the analyzer. If you use 100 percent O_2 you may have to wait a while for the reading to stabilize. This single-point calibration is usually sufficient, but if you wish you can also do a zero calibration with nitrogen. If you do so, flow the nitrogen through the analyzer for at least 10 min before adjusting the zero point. Always zero before spanning.

There is a curious tendency among newcomers to metabolic measurement to assume that a bottled O_2 span gas is more accurate than ambient O_2 levels after drying (chemical or mathematical; see Chapter 8) and BP correction. For systems operating far away from ambient O_2 levels this may be true, but for the clear majority of metabolic measurement systems this is a profoundly ignorant superstition. Outside air across the entire planet has an O_2 concentration of 20.94 percent after removal of water vapor and correction for BP (Tohjima et al., 2005). In a small ill-ventilated room containing a bustling group of people, CO_2 levels can rise to as much as ~ 2000 ppm or 0.2 percent, about 0.16 percent above background atmospheric levels. Taking accompanying O_2 depletion into account, it can be shown that this unusual and suboptimal circumstance may cause ambient O_2 levels to fall from 20.94 percent to ~ 20.91 percent. This is an error of ~ 0.14 percent if you assume a concentration of 20.94 percent when spanning your O_2 analyzer. In contrast, even the finest commercially available span gases are accurate to 1 percent at best. So, if you insist that your tank of O_2 calibration gas is superior to WVP- and BP-corrected ambient O_2 concentrations (even under the worst of unlikely conditions!), you have increased your span error by *at least* 700 percent. If, as required by most IACUC protocols and government regulations, your facility is ventilated by clean outside air and has a good HVAC system, the ambient O_2 error declines to ~ 10 ppm or 0.001 percent, making bottled O_2 span gas fully a *thousand* times less accurate! Plus, ambient air is free and won't run out; what's not to love?

CARBON DIOXIDE ANALYZERS

Carbon dioxide molecules contain C=O bonds that act like springs, causing the molecule to vibrate if hit by light at any of several wavelengths. One particularly effective wavelength is 4.26 μm in the mid-infrared, at which CO_2 absorbs energy, vibrates, and reemits the energy at a longer wavelength. Indeed, this is one basis of the greenhouse effect. All commonly available CO_2 analyzers make use of this property of CO_2 and are generically referred to as infrared gas analyzers (IRGAs). Light from an infrared emitter is absorbed by CO_2, and the extent of this absorbance allows the amount of CO_2 present in the optical path to be measured.

As anyone familiar with the Boltzmann distribution of emitted wavelengths of light versus the temperature of the radiating body can tell you, everything at room temperature glows weakly at 4.26 μm wavelength, so at first sight, detecting changes in infrared intensity caused by absorbance by CO_2 would seem like trying to take photographs with a glass camera. This problem is avoided by modulating the infrared source and by using various techniques to make the sensor sensitive only to changes in the level of infrared light impinging on it.

Most good CO_2 analyzers are explicitly or implicitly differential. In other words, they measure either the difference in CO_2 content between two optical paths (explicitly differential), or infrared absorbance through a single optical path at two different wavelengths (implicitly differential). These features help control for drift and aging in the source, detector, and associated electronics.

An optical filter centered around 4.26 μm limits the response of the optical bench to CO_2 specifically. An alternative approach uses a diffraction grating to create a pure source of 4.26 μm light; this type of light source is called dispersive. Filters are not dispersive, and so CO_2 analyzers that use filters are sometimes referred to as nondispersive infrared (NDIR) analyzers. Almost all IRGAs are of the NDIR type. The actual optical path through which the sample gas flows is usually made of polished metal, often gold-plated to resist corrosion. Obviously, any dirt or dust that enters this path will interfere with light transmission, so air entering CO_2 analyzers must always be carefully filtered unless an internal filter is present. Allowing air to enter through a CO_2 analyzer's OUT port will bypass any internal filtration, guaranteeing noise and drift problems. Because the optical path usually has a low resistance to flow and contains no delicate components (though it *does* have to be kept clean), CO_2 analyzers are usually not particularly sensitive to flow rates.

Because CO_2 analyzers measure upward deflections from a very low baseline (~ 0.04 percent CO_2 in background air, but in practice sometimes scrubbed to 0 percent CO_2), they are far more sensitive and drift far less than O_2 analyzers, which measure tiny deflections from a huge baseline, which in a crowded laboratory might not be stable enough for extremely critical small-animal work (e.g., small groups of *Drosophila*) unless care is taken to make sure it is (see Figure 4.2). CO_2 analyzers are ideal for measuring the MR of small organisms; the best CO_2 analyzers can easily resolve the CO_2 production of a single *Drosophila melanogaster* in a flow-through system (see, for example, Lighton, 2007; Lighton and Schilman, 2007).

Calibration of CO_2 analyzers

CO_2 analyzers require two gases for calibration. One, the zero gas, contains no CO_2, while the other, the span gas, contains a known concentration of CO_2, usually in nitrogen. Purists may wish to specify "air"—20.9 percent O_2, the balance nitrogen—as the carrier gas, because the carrier gas marginally affects the infrared absorption characteristics of CO_2. However, the effect is so small compared to other sources of error in most respirometry systems that the extra complexity and expense is generally not warranted. The zero gas is usually either nitrogen or air scrubbed of water vapor and CO_2 by scrubber chemicals such as Drierite and Ascarite. To calibrate a CO_2 analyzer, the zero gas is pushed or pulled through it and the analyzer's reading is adjusted to zero. The same procedure is followed with the span gas, except that the analyzer is adjusted to read the known concentration of the span gas. These procedures are repeated until no further adjustment is necessary.

All CO_2 analyzers are nonlinear, but most manage to hide the fact more or less successfully. The relation between the optical absorbance and concentration of substances that absorb light generally follows the Beer–Lambert law, which means that they show a linear relationship between concentration and optical absorbance. CO_2, not content with flagrantly disobeying the Beer–Lambert law, effectively farts in its general direction. Manufacturers of CO_2 analyzers can choose to correct for this nonlinearity via any of many proprietary techniques. The worst use low-resolution look-up tables or piecewise-linear techniques; the best calibrate each analyzer individually at many CO_2 concentrations and construct exact linearization functions for each analyzer they sell. Whatever the linearization technique, all linearized analyzers display CO_2 concentrations in final units, typically in percent or parts per million. A few older analyzers are not linearized at all and rely on their users to linearize their output. In such cases, the manufacturer usually supplies a linearization function, typically a third- or fourth-degree polynomial equation, to transform the voltage output of the analyzer into final concentration units. Thus, spanning the analyzer usually requires it to be connected to a data acquisition system.

It is necessary to be aware of this behind-the-scenes nonlinearity, which has two important effects even in nominally linearized CO_2 analyzers. First, to minimize the errors inherent in linearization, always span a CO_2 analyzer at a CO_2 concentration within the range of values it will likely measure. For example, most flow-through respirometry setups for vertebrates will, or at least should, yield excurrent CO_2 concentrations in the 0.1–0.5 percent range. A suitable span gas would be, say, 0.5 percent. Resist the temptation to borrow a respiratory physiologist's 5 percent CO_2 span gas, intended for the calibration of end-tidal CO_2 analyzers, or a botanist's 0.035 percent CO_2 span gas, intended for photosynthesis studies. Second, because the behind-the-scenes nonlinearity is worst at the highest CO_2 concentrations, take care to rezero CO_2 analyzers frequently, especially if you use them to measure high concentrations of CO_2. This is because minor drift close to zero, where the absorbance curve is shallow and almost linear, can translate to a significant error when that drift is shifted upward by the sample gas and affects CO_2 concentrations in a steeply nonlinear section of the absorbance curve. Always zero first, then span second.

Note that if a CO_2 analyzer has poor linearity—as many industrial models do, because they are not individually linearized—you may need to calibrate it at two concentrations of CO_2 rather than zero CO_2 and a single span concentration. This is because its linearized response cannot be trusted to extend down to 0 percent CO_2. Typically such analyzers must be calibrated at the lowest and highest CO_2 concentrations they are likely to encounter—in the case of metabolic phenotyping systems, usually 0.04 percent and 1 percent. This is a mark of questionable analyzer quality and will double the error margin, because not one but two gas mixtures with associated (in)accuracies will be used.

Because of all these factors, you would be wise to take claims by CO_2 analyzer manufacturers or resellers of accuracies better than 1 percent of reading (e.g., 0.04 percent or even 0.001 percent) with a mole of salt. This is especially so because a CO_2 analyzer cannot be more accurate than the span gas used to calibrate it, while the accuracy of most CO_2 span gases is questionable and in practice never better than 1 percent of reading. Be sure to check the cylinder's label for its actual, as opposed to nominal, CO_2 concentration. Even if the stated accuracy of the actual (not nominal) concentration is 2 percent or 5 percent, it is not atypical for the real value to be outside of these bounds. Some forms of trickery are also quite common, especially implying "of-reading" accuracy when absolute accuracy is quoted. For example, a well-known metabolic cart manufacturer states that the accuracy of its CO_2 span gas is 0.03 percent. This amazed me, as it exceeds even National Institute of Standards and Technology (NIST) gold-standard, hyperexpensive calibration gas accuracies by an order of magnitude. So, I called them up and drilled down to the actual specifications. It turns out that the value of the span gas can be between 0.97 and 1.03 percent—an absolute accuracy of 0.03 percent, indeed, but a 3 percent accuracy specification in any meaningful, percent-of-reading sense of the term!

As a rule of thumb, any analyzer or metabolic phenotyping system manufacturer claiming a CO_2 analyzer accuracy better than 1 percent of reading is ignorant, dishonest, or both.

Some workers carefully check the RQ of a gas flame (Chapter 14) and adjust the CO_2 span gas's concentration accordingly, but of course this requires great faith in the functionality of the O_2 measurement system.

One minor point remains. The behavior of CO_2 in the infrared is affected by the other gas species with which it shares the sample, such as N_2O which has a major effect. Water vapor very slightly weakens CO_2 infrared absorbance, a phenomenon known as line broadening that reduces the peak magnitude of the CO_2 absorbance curve. Thus, measuring CO_2 in the presence of water vapor may lead to a tiny underestimate of CO_2 concentration, quite apart from the dilution effect of water vapor (equation 8.7). My experience with systems that utilize mathematical compensation for water vapor dilution suggests that this effect can be ignored; they yield RQ values within experimental error of those predicted by flame stoichiometry and the FQ = RQ equivalence (see Chapter 14).

WATER VAPOR ANALYZERS

Water vapor in air can be measured using three distinct technologies, which have characteristic strengths and weaknesses. The ultimate point of any water vapor measurement in respirometry is to obtain the partial pressure of water vapor in the air, and therefore the most user-friendly analyzers are the tiny minority that provide a direct readout of this variable. Most water vapor analyzers that do not provide a direct readout of water vapor pressure do, however, provide readouts that can be converted to water vapor pressure, directly or indirectly. In addition, the ability to analyze water vapor in-line (in a flow-through system, as opposed to waving the sensor about on a stick) is almost essential.

Chilled mirror water vapor analyzers

Chilled mirror water vapor analyzers are the most accurate. They measure water vapor pressure from the dewpoint of the air, using a chilled mirror that is held at the point where water vapor in the air starts to condense or freeze. The dewpoint of air is a fundamental function of its water vapor content. Because the temperature of the mirror can be measured accurately (from $\pm 0.5°C$ to $\pm 0.1°C$ in the best analyzers), chilled mirror analyzers are outstandingly accurate and do not require calibration other than for their temperature measurement component. Chilled mirror analyzers are almost invariably of the in-line type, which is convenient, but care needs to be taken not to allow any particulate material to enter them and contaminate the mirror. From the measured dewpoint, the partial pressure of water vapor can be calculated using a variety of nonlinear mathematical functions (see http://www.respirometry.org for an online calculator).

Though much admired for their accuracy, chilled mirror analyzers are less than ideal laboratory companions. They are slow-responding, bulky, expensive, and generally do not have the ability to convert dewpoint to water vapor pressure; you will usually have to calculate the water vapor pressure of the air stream yourself (equation 16.7).

Capacitive water vapor analyzers

Capacitive water vapor analyzers are the most common variety and the least expensive. They rely on water vapor adsorption by a sensor, which results in a change in the sensor's capacitance. This change in capacitance is usually converted to a change in the frequency of an oscillator, which in turn is converted to a voltage. Because in most cases the sensor's adsorption of water vapor is proportional to the saturated water vapor pressure at the temperature of measurement, they directly measure relative humidity (RH), which is simply

$$RH = 100(WVP/SWVP) \qquad (16.5)$$

where WVP is the water vapor pressure and SWVP is the saturated water vapor pressure at the temperature of measurement. Thus, we can easily obtain water vapor pressure:

$$WVP = (RH/100)SWVP \qquad (16.6)$$

Now we need to find the SWVP, which is a function of temperature. Hence the temperature at which the RH measurement was made must be known. A suitable equation for calculating SWVP from temperature is:

$$SWVP = (0.61121)e^{(18.678 - T/234.5)[T/(257.14+T)]} \qquad (16.7)$$

where SWVP is in kilopascals and T is the temperature of the RH sensor in degrees Celsius (Buck, 1981).

Capacitive water vapor analyzers do have some drawbacks. Most are not capable of in-line measurements without modifications; most do not provide water vapor pressure readouts directly; and all require regular calibration in accordance with the manufacturer's instructions or your own ingenuity. One unorthodox water vapor analyzer calibration technique makes use of Nafion tubing (see Chapter 19), which is permeable to water vapor but not to O_2 or CO_2. The Nafion tubing is immersed in distilled water, and the air pushed through it takes on the dewpoint of the water surrounding it, which is equivalent to the water's temperature and can be measured accurately. A water vapor analyzer can

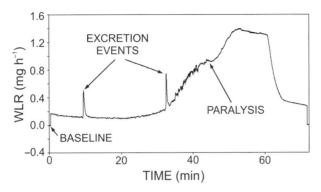

Figure 16.2 Water loss from a single *Drosophila melanogaster*, mass about 1.2 mg, as its temperature is ramped from 30 to 45°C. The trace was produced using an in-line water vapor analyzer (Sable Systems RH-300) with direct readout in micrograms per milliliter of water vapor density. Raw data are from Lighton (2007).

be calibrated accurately against this known dewpoint or the calculated water vapor pressure to which it corresponds (equation 16.7).

On the plus side, capacitive water vapor analyzers are accurate (typically 1–2 percent RH) if well calibrated, and they are lightweight, robust, and inexpensive. Good instruments are easily capable of measuring over a wide range of water vapor pressure, and some, apart from delivering a wide range of measurements, have such high resolution that they can easily provide accurate water loss information from a single *Drosophila* (Figure 16.2).

Infrared water vapor analyzers

Infrared water vapor analyzers are often part-and-parcel of high-quality CO_2 analyzers. To achieve this, the infrared beam is split with a pellicle, and half is dedicated to measuring the infrared absorption of water vapor at a wavelength of 2.59 μm. The result is an analyzer that reads out water vapor pressure in parts per thousand at ambient pressure. Thus,

$$\text{WVP} = \left(\text{BP} \times \text{PPT}\right) / 1000 \tag{16.8}$$

where BP is barometric pressure in kilopascals and PPT is the analyzer's reading in parts per thousand. As with CO_2, water vapor does not obey the Beer–Lambert law, so the output of the analyzer may need to be linearized in accordance with the manufacturer's instructions. Although not as high resolution as the best capacitive water vapor analyzers, infrared water vapor analyzers are excellent instruments, though extremely expensive. Probably their largest single advantage, in the case of dual-wavelength analyzers, is that they yield water vapor data that are almost perfectly temporally correlated with CO_2 data. Note "almost perfectly," not perfectly, because water vapor is more prone than CO_2 to adsorbing onto surfaces such as tubing, which tends to blunt its response times. This effect can be minimized by using tubing made of or lined with highly hydrophobic materials such as polytetrafluroethylene (PTFE) (see Chapter 19).

Calibration of water vapor analyzers

The reader is referred to Chapters 8 and 4 (especially Figure 4.10 and associated text) plus the discussion in the section "Capacitive water vapor analyzers," concerning Nafion tubing for more on this topic.

17

The Varieties of Flow Meters

Why do we emphasize measuring flow? The answer is simply because flow-through respirometry has an error budget that most of us would like to minimize, and in typical systems, the measurement of air flow rates is the largest single contributor to that error budget.

Of the various technologies for measuring flow rates of air, by far the most commonly used in the laboratory are rotameters and mass flow meters. The mass flow meter, which has an electrical output and so can provide information that other electrical circuitry can make sense of and that can be recorded continuously, can be linked in a control loop to a gas flow control element such as a variable-aperture solenoid or a variable-drive pump to create a flow controller. A third class of flow measurement methods exists primarily for calibrating and verifying the correct operation of the others; these will be dealt with in the section on nitrogen dilution calibrations at the end of the chapter.

ROTAMETERS

Rotameters are the simplest flow meters. They consist of a very slightly tapering bore that passes through a glass or acrylic body and contains a float (typically a ball made of glass or steel). As the rate of air flow entering at the bottom of the bore increases, more force is imparted to the float, which levitates to the level where the force of the air pushing upward on it (and escaping through the annulus of space between the tapered bore and the float) equals the force of gravity on the float. There the float sits; it can do no other. If a graduated scale is placed alongside the bore, typically etched, stamped, or printed in place, the vertical displacement of the float can be quantified. Rotameters typically come with a scale graduated in flow units or in arbitrary units that are translated into flow with a look-up table supplied with the meter. In either case, the manufacturer should specify the conditions (barometric pressure and temperature) under which the measurements were made.

The measurement conditions under which the rotameter was calibrated are important because the rotameter is an intrinsically volumetric flow meter. It measures volumes of flow per unit time at ambient temperature and pressure, rather than the molar quantity

Measuring Metabolic Rates: A Manual for Scientists. Second Edition. John R. B. Lighton, Oxford University Press (2019).
© John R. B. Lighton (2019). DOI: 10.1093/oso/9780198830399.001.0001

of gas passing through it. In measuring metabolism, we are not specifically interested in arbitrary volumes of O_2 consumed or CO_2 produced; we need to know how much was produced or consumed in molar terms. This requires that volumetric flow rates be converted to molar flow rates by correcting the measured volume to standard temperature and pressure:

$$STPFR = FR(BP/BPS)(TS/T) \tag{17.1}$$

where STPFR is the flow rate corrected to STP, FR is the uncorrected flow rate, BP is measured barometric pressure, BPS is standard barometric pressure (e.g., 101.325 kPa or 760 Torr), all in the same units; and T is temperature and TS is standard temperature (273.15 K), both in Kelvin. If you are using a rotameter calibration with a specific non-STP calibration temperature and pressure, you will first need to convert your readings to the calibration temperature and pressure using equation 17.1. To do this, replace the STP in the equation 17.1 with the calibration temperature and pressure. Finally, convert the result to STP. Some rotameter manufacturers recommend their own proprietary equations, which should be used if they are specified.

It's certainly a good idea to keep track of ambient barometric pressure and temperature, but having to correct flow rates in this fashion quickly becomes tedious. It is also inaccurate. Unless rotameter readings, rotameter temperature, and barometric pressure are continually recorded (which is not easy because rotameters are intrinsically analog and visual, not electrical), dynamic corrections cannot be made, and flow drift can be quite significant over the course of a recording. This is especially the case if (apart from pressure and temperature fluctuations) the pump drive is not constant or if the flow system's resistance changes. Add to this the fact that there is no practical method of recording the rotameter's reading as part of a conventional data stream, and you can see that the rotameter is an instrument of last resort. It is seldom used anymore for research. For teaching it is adequate, where its relatively low cost and obvious mode of action are even advantageous.

MASS FLOW METERS

Mass flow meters measure, as the name implies, the molar mass of air passing through them. Very few explicitly measure flow rates in moles per unit time; rather, the common convention is to express the flow rates in STP-corrected volumes. Because of their principle of operation (see below), they are basically unaffected by fluctuations in temperature or pressure and will placidly read STP-corrected flow rates while ambient conditions fluctuate. Moreover, mass flow meters all have an analog or digital output that can be recorded along with other data. Conventionally, this analog output varies from 0 to 5 V as the flow through the meter varies from 0 to 100 percent of full scale. Thus, for example, an output of 1.00 V corresponds to a flow rate of 20 percent of full scale.

There is one peculiarity to be aware of: Although practically all mass flow meters are calibrated against sea-level *pressure* (760 torr or 101.325 kPa), many mass flow meters are calibrated to a default "room temperature" rather than true standard temperature (0°C or 273.15 K). These meters are primarily intended for process control, and their actual calibration temperature will be listed on their calibration label, which is stuck to the sensor body. It is often 73°F, which equates to 22.77°C. Thus, the signal from these meters should be multiplied by 0.923 [273.15/(273.15 + 22.77)] to yield true STP flow rates. At time of purchase, or when having the meter recalibrated, you can always request a true-STP calibration; the manufacturer might even oblige, but this should be verified by scrutinizing the calibration label, instead of being assumed.

Most mass flow meters work by heating air and measuring the ability of the heated air to transfer heat elsewhere at a rate depending on flow rate. These are referred to as thermal mass flow meters, and they can be intrinsically quite linear. Variants abound, some of which (especially "microbridge" mass flow meters, which use micromachined, microminiature heat transfer sensors) can be quite nonlinear. Almost all commercial mass flow meters incorporate a linearity correction of some kind, which varies in complexity from a potentiometer or two to a digital data link that allows calibration coefficients to be burned into reprogrammable nonvolatile memory. A few extremely cheap meters give a nonlinear (and not very stable) output and expect their users to recalibrate them at frequent intervals, but this is not the case with products from reputable manufacturers. Most thermal mass flow meters can be fairly inexpensively reranged by swapping out a laminar flow resistance element and recalibrating them. These operations, like the calibration of the more sophisticated flow meters, are best left to the manufacturer.

An alternative and increasingly common variety of mass flow meter measures the differential pressure across a restriction. This pressure varies according to the square root of the flow rate and is usually linearized by the meter's electronic innards. If the flow is carefully engineered to be as laminar as possible, such as in those made by the Alicat company (Appendix 1), the precision of linearization is quite impressive.

With either form of mass flow meter, you can always increase its upper range limit by bypassing it, using a T-junction and a second limb of tubing. This will, of course, totally alter its calibration, requiring you to calibrate it in its new configuration.

Most mass flow metering or controlling valves are specified to an accuracy of 1–2 percent of full scale. They are usually recommended for operation down to about 10 percent of full-scale flow rate, also known as a turn-down ratio of 10:1. Thus, a meter capable of giving 1 percent accuracy at, say, 1000 mL min^{-1} has an accuracy at 100 mL min^{-1} of (1 percent)(1000/100), or 10 percent of reading. This may be less than ideal for your application, especially if you are using mass flow meters for creating an accurate mixture of gases. Be aware of this quirk, and do not use conventional mass flow meters at flow rates below the point where their accuracy deteriorates to levels you consider unacceptable in your application. This is the reason that most scientists who use these meters soon acquire a menagerie of them with overlapping ranges, such as 10–100,

20–200, 50–500, and 100–1000 mL min^{-1}. As alluded to above, any mass flow meter can be plugged into a control loop with an effector such as a variable-orifice valve. The resulting mass flow control valve usually needs an electronic interface to deliver its supply and control voltages, although some manufacturers have started to produce mass flow control valves with intrinsic settability and readouts (Alicat, again). The basic principle is simple: The control unit delivers a voltage (usually in the range 0–5 V) to the mass flow control valve, in addition to the fixed voltages required to make it work (usually a bipolar 15-V or single 24-V DC supply). The mass flow control valve then regulates the flow passing through it to the corresponding proportion of its maximum flow rate. For example, setting the control voltage of a 1000-mL min^{-1} maximum mass flow control valve to 1.00 V will cause it to regulate at 1000 (1/5), or 200 mL min^{-1}.

Mass flow control valves have the same accuracy issues as described above for mass flow meters in most cases. Exceptions to this are the control valves made by the Sensirion corporation, which are accurate to 0.8 percent of reading from 10–100 percent of full scale, and those made by the Alicat corporation, which are available with accuracy as impressive as 0.4 percent of reading (at extra cost) and turn-down ratios down to 100:1. The Sensirion valves require a very high differential pressure (at least 150 kPa) to regulate throughout their range, but the Alicat valves are much more forgiving. Most other mass flow control valves also require quite high differential pressures to operate. For use in respirometry, I suggest having the manufacturer calibrate your valves at the lowest available differential pressure, which is typically about 30 kPa. This is not an option with the Sensirion valves, which use a low-travel piezoelectric actuator that requires high differential pressures for controlling flow. Note that the differential pressure does not affect the metering of flow, only the ability of the variable flow element (typically a variable-orifice solenoid) to maintain control in the closed-control loop.

All thermal mass flow meters and mass flow controllers are affected by the thermal conductivity of the gas passing through them. Most are calibrated for air or nitrogen. The calibration gas is always specified at the time of purchase and on the unit's label. The only common gases used in respirometry that present problems in this respect are CO_2 and helium. Both are significantly more thermally conductive than air, causing thermal mass flow meters (and thus thermal mass flow control valves) to be more sensitive, and therefore to overestimate readings. Thermal mass flow meters and controllers are usually supplied with tables of k-factors for correcting flow rates to particular gases. For example, pure CO_2 has a k-factor of 0.7; a stream of pure CO_2 passing through a flow meter calibrated for air or nitrogen would read 100 mL min^{-1}, while the actual flow rate is a mere 100(0.7), or 70 mL min^{-1}. In mask applications, it may be necessary on occasion to compensate for this effect if excurrent CO_2 levels are very high. However, even under the worst of circumstances, sustained excurrent CO_2 concentrations are unlikely to exceed 1 percent from a mask pulled at a respectable flow rate (see Chapter 12), making the maximum error from this source approximately 0.3 percent, which is insignificant relative to the total error budget of a typical respirometry system. Other mass flow meters and mass flow controllers, such as the Alicat range, operate by measuring the differential pressure generated across a laminar flow element. In such devices the gas density, rather than its thermal conductivity, determines the calibration of the meter. Tables of gas densities are available from the manufacturer. Flow meters and flow controllers of this type can be set to a particular gas over a serial link, or even via a front panel keypad.

CALIBRATING FLOW METERS

Calibrating mass flow meters (and mass flow controllers) can be complicated and is best left to the manufacturer. However, some relatively simple techniques exist for measuring flow rates accurately, and these are useful for checking a meter's calibration or even creating a correction calibration, whereby the meter's output is corrected to the actual flow rate by a corrective equation or look-up table. The methods described below are impractical for routine measurement of flow rate; they are intended mainly for checking the flow rate readings of another meter. Given the importance of accurate flow measurement to flow-through respirometry, we are treating flow calibrations in some detail, especially those that rely on nitrogen injection. Those techniques are principally used with high flow-rate systems, especially mask-type systems that operate at high flow rates (see Wickler et al., 2003, for a good example). At lower flow rates, bubble flow meters are commonly used and will be described first.

Bubble flow meters

Bubble flow meters and their dry cousins (typically referred to as piston flow meters) are charmingly intuitive. A soap bubble is chased up a precision-bore tube by the flow, and its time of passage between two points is accurately timed. For the basic operating principles, see the classic paper by Levy (1964). Basically, the cross-sectional area of the tube is multiplied by the distance between the two points in the same units, yielding a volume. The volume is then divided by the passage of time (usually measured in seconds, then converted to minutes). It's possible to do this with a meter stick and stopwatch, but most commercial bubble flow meters use electronic timers based on optical gates. Such meters will also give a volumetric flow rate reading directly, based on the tubing size installed. It is then straightforward to convert the volumetric reading to STP, using equation 17.1. Some of the fancier units will do this for you. The only slight complication is that the reading should, for best accuracy, be compensated for the addition of water vapor pressure to the total partial pressure mix (equation 8.6). This, in turn, requires a look-up table of water vapor pressure versus temperature (or an equation relating the two; see equation 16.7), and it also requires local barometric pressure measurement (as does STP correction). Dry piston flow meters don't require water vapor correction, of course, but inevitably suffer from higher friction than a soap bubble. Bubble flow meters (especially the Gilian Gilibrator, now made by the Sensidyne company of Clearwater,

FL, USA) are widely used in air-monitoring applications and are often available at low prices from Internet auction houses. They are a useful addition to any respirometrist's armamentarium.

Nitrogen dilution method

The nitrogen dilution method is an alternative method for calibrating mass or volumetric flow meters, especially at higher flow rates. It requires an accurate oxygen analyzer, which most people have or can readily obtain. Proceed to start the flow you want to calibrate. Pull a subsample of the flow (downstream from the flow meter you're trying to calibrate if that's your intention) through the oxygen analyzer. About 200 mL min^{-1} will work well. *Do not scrub anything from the main or the subsampled flow unless directed otherwise.* The plan is to quantify the main flow rate by observing the dilution of the oxygen within the main flow caused by the controlled addition of an oxygen-free gas, traditionally nitrogen, to the flow.

At this point you can decide: whether to add the nitrogen to the intake or the outlet of the flow-generating system. Obviously, if you add nitrogen to the intake, you must do so at ambient pressure (i.e., without either forcing it in or impeding the flow of air). I first describe in some detail using the nitrogen dilution technique by adding nitrogen *after* the flow meter. Next, I turn to using the nitrogen dilution technique by adding nitrogen *before* the flow meter. If you are using the latter technique, you should still read through both descriptions to get an adequate background. It can certainly be argued that adding nitrogen before the flow meter is a more realistic technique, and it has the advantage that in most systems the nitrogen will be more thoroughly mixed into the air by the time the flow is subsampled.

Postflow meter nitrogen dilution

The addition of nitrogen should always take place *downstream* from the meter, if any, for obvious reasons, and care should be taken to ensure adequate mixing of the nitrogen with the air stream by adding some baffles to induce turbulence downstream from the injection source, for example, before a subsample is withdrawn for analysis.

At this point you need to know three quantities accurately: the concentration of O_2 in the ambient air, the flow rate of nitrogen into the main flow downstream from the flow meter (if any), and the exact O_2 concentration of the nitrogen-diluted air.

The flow rate of nitrogen injected into the system is critical. A good, fairly recently calibrated or validated mass flow controller should be used. If a mass flow meter is being calibrated for the main flow, be sure to check that the calibration temperature of the injection mass flow controller matches that of the mass flow meter. If the injection mass flow controller is calibrated to "73F" (i.e., 73°F or its decimal system equivalent) then its reading should be multiplied by 0.923 if the main flow meter is being calibrated to strict STP. If a volumetric as opposed to a mass flow meter is being calibrated for the main flow, the flow rate of the injection mass flow controller should, for the purposes of this exercise, be converted to volumetric flow:

$$VFR = STPFR(BPS/BP)(T/TS) \qquad (17.2)$$

where VFR is volumetric flow rate, STPFR is flow rate at STP as given by the mass flow controller, and corrected, if necessary, to standard temperature, and other variables are as before. This equation is basically equation 17.1 in reverse.

If you are using a volumetric flow meter for adding the nitrogen diluent, the exercise is hardly worth doing because the accuracy of most rotameters is so poor. However, it's better than nothing.

In short, you must keep track of whether you're calibrating a mass or a volumetric flow rate for the main flow, of the kind of injection flow meter or controller being used, and of the calibration temperatures and pressures of the injection and the main (if any) flow meters or controllers. Mixing incompatible flow measurements will cause untold confusion.

Next, we consider ambient oxygen levels. It's certainly possible to measure relative humidity and ambient temperature and calculate water vapor partial pressure (or, if you have a smart water vapor analyzer, to measure water vapor partial pressure directly). Then, together with knowledge of ambient barometric pressure, you can calculate the dilution of nominally 20.94 percent dry atmospheric O_2 and span the O_2 analyzer to this figure, provided that you are reasonably certain that the figure is accurate in your situation (ideally, outdoors or in a very well-ventilated room). The equation for doing this is intuitive:

$$O_2d = (O_2 \times BP)/(BP - WVP) \qquad (17.3)$$

where O_2d is the dilution-compensated (dry) O_2 concentration, WVP is water vapor pressure, and BP is barometric pressure in the same units. A much easier, faster, and arguably more accurate way to make this measurement is to temporarily place a H_2O–CO_2 scrubber (Drierite/Ascarite/Drierite, for example) before the O_2 analyzer, span it to 20.94 percent (if applicable in your situation; see Tohjima et al., 2005 if you are operating with ambient air, in which case 20.94 percent is certainly applicable), and then remove the scrubber: You are now measuring the O_2 concentration of water-vapor-diluted, CO_2-diluted air without having to do any calculations or make any other measurements or assumptions. You will note that the new reading is significantly lower than the spanned O_2 reading on dry, CO_2-free air.

With your calibrated O_2 analyzer taking readings, inject the nitrogen at a known flow rate compatible with the flow rate you want to measure: volumetric if volumetric, mass if mass. Now, it is obvious that

$$O_2d = O_2i(FR)/(FR + NFR) \qquad (17.4)$$

where O_2i is the initial concentration of O_2 and O_2d is the nitrogen-diluted concentration of O_2 in the same units, with nitrogen being added to the main flow rate, FR, at rate NFR, also in the same units. From this it follows that

$$O_2i = O_2d + O_2d(NFR / FR) \qquad (17.5)$$

and rearranging further,

$$1 / FR = (O_2i - O_2d) / (O_2d)(NFR) \qquad (17.6)$$

and thus

$$FR = O_2d(NFR) / (O_2i - O_2d) \qquad (17.7)$$

Obviously, this equation approaches a singularity when $O_2i = O_2d$, so NFR needs to be a reasonably significant fraction of FR—certainly not less than about 1 percent. Multiple readings should be taken and averaged.

As a typical example, assume that you are injecting an NFR of 5 L min^{-1}, and the O_2 reading falls from 20.802 percent (remember that the O_2 is diluted by water vapor and CO_2) to 20.309 percent. The principal flow rate is

$$(5.0 \times 20.309) / (20.802 - 20.309) = 101.545 / 0.493$$

or 206 L min^{-1}.

Water vapor correction is a source of much confusion, especially in pull systems. If you are pulling or pushing dry air through a system, you don't need to be concerned with the effect of water vapor on flow rates unless you are measuring flow rates downstream from an animal that is adding water vapor to the flow. If, however, you are pulling air at ambient humidity levels from the environment and if you are drying the air before gas analysis, you will need to compensate for the presence of water vapor as shown in equation 8.6. But this compensation in no way affects the nitrogen dilution flow calibration technique; your mission is to find the total flow, water vapor and all, which is what a flow meter attached to the inlet of the system would measure. See the end of the next section for dealing with the small disparity in air water vapor levels (if any) that results from adding a flow of dry gas.

Preflow meter nitrogen dilution

The preflow meter technique was described in some detail by Fedak and co-workers in 1981, who primarily used it to calibrate measures of O_2 consumption. The principle is very simple. If you add pure nitrogen (or any other oxygen-free gas, such as argon) at flow rate NFR to the input of a flow-through respirometry system *without changing its main flow rate*, and if the incurrent fractional concentration of O_2 is F_iO_2, the system will behave as if O_2 is being withdrawn at a rate of $NFR(F_iO_2)$. For present purposes, we are using the technique only to measure flow rates and so do not need to go to the level of detail of the original article (Fedak et al., 1981), to which you should refer for a more detailed treatment. In particular, we will simplify the treatment by not allowing for the fact that we are adding dry nitrogen to an air stream that may contain water vapor; Fedak et al. could not make this simplifying assumption because they treated the analysis of dried gases only and thus had to correct the main flow rate to remove the water vapor contribution (equation 8.6).

The O_2 analyzer is spanned as described above, using a water vapor and CO_2 scrubber, to an oxygen concentration that is reasonable for the environment—typically 20.94 percent for outside air (Tohjima et al., 2005). A subsample of the excurrent flow is analyzed without drying. Unlike the case with postflow meter mixing, adding more nitrogen to the inlet rapidly displaces O_2 in air, until, when nitrogen flow rate equals the main flow rate, no more O_2 is theoretically present in the main flow. This is in contrast to the technique above, where the O_2 concentration in the main flow was halved when NFR equaled FR. Thus, in this system,

$$O_2d = O_2i(FR - NFR)/FR \qquad (17.8)$$

where units are as before. Following a similar chain of logic as above, this simplifies to:

$$FR = (O_2i)(NFR)/(O_2i - O_2d) \qquad (17.9)$$

You will notice that where FR = NFR, O_2d is forced to zero, whereas in the postflow meter case, O_2d is forced to $O_2i/2$. If NFR exceeds FR, nothing happens because the system flow rate remains at FR, and the excess nitrogen simply spills into the room, possibly discommoding passersby. Note that as with the postflow meter injection technique, we are neglecting the fact that the main flow may contain air at ambient water vapor pressure, while the nitrogen will be dry. At an NFR of < 5 percent of FR, this effect will be negligible unless you are sampling air from a tropical swamp. Should you want to compensate for it nevertheless, simply dry the subsampled air stream and correct FR for water vapor as shown in equation 8.6 before plugging it into equation 17.9.

To give a concrete example, suppose that we are injecting 5.0 L min^{-1} of nitrogen into the inlet of a flow system and that the O_2 reading falls from 20.802 percent (remember that the O_2 is diluted by water vapor and CO_2) to 20.309 percent, as in the first example. The principal flow rate is

$$(20.802)(5.0)/(20.802 - 20.309) = 104.01/0.493$$

or 211 L min^{-1}. Note that the preflow meter injection technique gives a different result under identical conditions of O_2i, O_2d, and NFR than the postflow meter injection technique. The disparity between the two techniques enlarges rapidly as NFR increases relative to FR.

Because you are injecting dry gas into a system, you will change the water vapor content of the system. This is obviously not a problem with dry air, but with moist air, as in a typical mask system that draws on ambient air, the water vapor content will be reduced by the amount (FR − NFR)/FR. If you are not scrubbing water vapor before O_2 analysis, you should account for the accompanying reduction in O_2 dilution by water vapor. In such a case, the easiest approach (if the error is potentially significant in your case) would be to measure water vapor pressure in the flow and dynamically correct the flow rate to remove the effect of water vapor as described (equation. 8.6). In this case, the air should be dried before O_2 analysis. An alternative approach is to humidify the nitrogen. One rather costly approach would be to flow the nitrogen through a section of parallel Nafion tubing (Chapter 19) sufficiently long enough to ensure that the nitrogen flow reaches water vapor equilibrium with ambient air.

If you want to use nitrogen dilution to calibrate oxygen consumption measurements along the lines of the Fedak et al. (1981) approach, see Chapter 14.

18

The Varieties of Activity Detectors

For those who study life forms that have the power to move, quantifying the metabolic rate of such an organism in a motionless state can be tricky and frustrating. This chapter describes various objective techniques for obtaining a qualitative or even quantitative index of activity that can be recorded along with other respirometry parameters. Though intended chiefly for users of flow-through respirometry setups, the principles apply to other systems as well, though with some cautions. For example, obtaining activity data during constant volume (also known as stop-flow) respirometry requires that all the chambers be monitored all the time, whereas with flow-through respirometry, only the chamber currently selected for recording needs to be monitored; all the other activity detectors can safely be ignored.

There are seven basic types of activity detectors. Commercial implementations are available for some, but not all. Most are easy to incorporate into voltage-based data acquisition systems; one (video) is not quite so straightforward.

OPTICAL ACTIVITY DETECTION

Optical activity detection, in its simplest form, gives a useful qualitative index of the activity level of almost any organism of any size. The idea is simple. When an organism is motionless, a light-sensitive detector (such as a photocell, photodiode, phototransistor, passive infrared detector, or light-dependent resistor) pointed in its direction will register no change beyond random noise. When the organism moves, the detector's voltage output will fluctuate (see Figure 7.2). This is easily recorded with any data acquisition system.

Obviously, a source of light is needed unless a passive infrared detector is used. This can be ambient light or any form of artificially generated light. Light-emitting diodes (LEDs) are a common choice. The light they produce can be of any spectral mix (near-ultraviolet, blue, green, orange, red, and white are common) or they can be invisible (880 nm in the near-infrared is typical).

The one small challenge is to create an activity data channel that is easily interpreted. The information lies chiefly in the fluctuations of the recorded signal, not its absolute magnitude. One approach to extracting this information is to take the first derivative of the activity channel. The result is a trace that is noisy when the animal is active.

Measuring Metabolic Rates: A Manual for Scientists. Second Edition. John R. B. Lighton, Oxford University Press (2019).
© John R. B. Lighton (2019). DOI: 10.1093/oso/9780198830399.001.0001

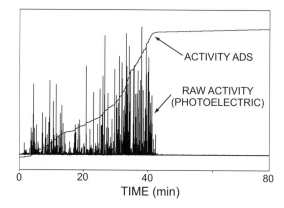

Figure 18.1 Photoelectric activity trace from a harvester ant, *Pogonomyrmex rugosus*, undergoing thermal stress. The raw data and the absolute difference sum (ADS) of the activity data are shown (units are arbitrary and are not shown). The breakpoint of the ADS is plainly visible and corresponds to the point at which the ant underwent heat paralysis. Raw data are from Lighton and Turner (2004).

Alternatively, the first derivative can be squared, eliminating the sign of the derivative. This makes it easier to quantify activity and to scan an activity channel during data analysis and determine the lowest activity level over a window lasting, say, 15 min. The corresponding section of the metabolic rate channel can then be assumed, on objective grounds, to be the least affected by activity metabolism.

An alternative approach is to sum the difference between successive data points in the activity channel, without regard to sign. The resulting absolute difference sum (Lighton and Turner, 2004) shows the steepest slope in the area of greatest activity and is at least semi-quantitative. It gives a clean breakpoint when activity ceases. Figure 18.1 shows a typical approach.

Alternative implementations of optical activity detection use linear or rectangular arrays of light gates (e.g., Lighton and Duncan, 1995) and give accurate position information on single organisms or positional activity information on multiple organisms. Commonly referred to as XY or XYZ arrays (the Z axis is for detecting rearing or standing behavior), these light gate arrays are widely used for quantifying the position and activity levels (the latter usually expressed as beam-breaks) of rats and mice in biomedical research. They are described in some detail in Chapter 13.

VIDEO ACTIVITY DETECTION

The technique of video activity detection requires a video camera, which in the simplest implementation can be as basic as a webcam, some of which give surprisingly good images. Some software packages intended for video analysis provide the capability for recording the average intensity of a selectable area of the image in real time, in which case you have a "smart" version of the optical activity detection described above. The resulting data are analyzed as for optical activity detection. The challenge comes in importing those data into your data acquisition system, preferably in real time. In some cases, it may be necessary to record the video data in a time-synchronized, separate file that is imported during data analysis. Video analysis is, of course, capable of far greater sophistication than described here; I am concentrating solely on objective activity detection.

MECHANICAL ACTIVITY DETECTION

Mechanical is used in its broad sense to mean mechanical forces generated by the animal. In its simplest form, a mechanical activity detector may monitor a respirometry chamber and detect changes in its center of gravity, for example by comparing the chamber's weight at three different points. As the animal within the chamber shifts position, the distribution of weight changes, and this can be measured with strain gauges or other piezoresistive sensors. The changes in all sensors are summed, and a voltage output proportional to differential motion is produced that can be easily recorded. The resulting data are basically very similar to the beam-break data obtained from optical activity detection and are analyzed in the same way.

A suitably configured piezoelectric activity detector can produce results that are qualitatively equivalent to those produced by a much more expensive quantitative optical beam-break system (but obviously, it is lacking position information, which can be important in some research projects)

MAGNETIC ACTIVITY DETECTION

Hall effect sensors detect magnetic fields, and linear output Hall effect sensors are very sensitive to changes in the position of a magnet in their vicinity. If a small magnet or a fragment of a magnet is stuck to an animal, the Hall effect sensor will detect any movement by the magnet, creating a voltage trace analogous to the above methods that is analyzable in the same way. Linear Hall effect sensors and their more sensitive cousins, giant magnetostrictive sensors, are available from electronic distributors (see Appendix 1). Some information on applying them can be found in Carlson (1996).

CAPACITIVE ACTIVITY DETECTION

All animals have a substantial water content, and water has a high dielectric constant, so any movement of an animal will change the electrical capacitance of its surroundings. This can be measured by setting up an alternating electrical field between two plates or wires near or to either side of the animal. As the capacitance between the two electrodes changes, the complex impedance caused by the capacitance between the electrodes also changes, and this change can be sensed by appropriate circuitry. Generally, capacitive sensing is restricted to very short ranges; this can be an advantage in some cases. A web search for capacitive sensors will yield useful information if you want to explore further.

PASSIVE INFRARED ACTIVITY DETECTION

If an animal's temperature differs from that of its surroundings, it will create fluctuations in the mid-infrared region of about 8–20 μm wavelength, where its Boltzmann radiation is significant. These fluctuations can be detected with a pyroelectric detector, which is typically a lithium tantalate crystal behind a germanium or silicon window (transparent to mid-infrared radiation) and housed in a small metal can about the size of a pencil eraser. Pyroelectric detectors are available in amplified and unamplified versions and are sensitive only to fluctuations in infrared level, not to the absolute magnitude of impinging infrared radiation, so they have already done your differentiation for you. To get reasonable range, an infrared focusing system of some sort is usually needed. This is typically a plastic Fresnel lens (available from Edmund Scientific, though usually supplied mounted to pyroelectric detectors intended for this application). Suitable pyroelectric detectors can be purchased from many sources on the web. It is also possible to use a security-alarm-type sensor with a switched output; the switched output is set to the test (lowest duration) position and fed into a low-pass filter with a time period of about 60 sec before being recorded. The result is a voltage signal that is somewhat proportional to the number of times the motion detector "tripped" in the last minute or so.

RADAR ACTIVITY DETECTION

Radar activity detection is, of course, chiefly an option for larger animals at longer distances, and even through light-opaque surfaces. Normal precautions regarding exposure to microwave radiation should be observed. If you examine the inside of a standard radar security sensor, you will find a microwave horn or pair of horns. At one focus is a Gunn diode, which produces microwave radiation that streams from the horn when activated by an applied voltage. At another focus is a Schottky detector/mixer diode that produces a voltage proportional to the level of microwave radiation that is reflected back into the unit from outside sources. Complete subunits are readily available on the web (search for "Gunn source") and will provide an analog output that can be directly recorded. The resulting data are similar to those obtained via optical activity detection and are analyzed in the same way.

19

The Varieties of Scrubbers, Tubing, and Tubing Connectors

PRESCRUBBING AIR

For many respirometry applications involving very small animals, it is helpful to scrub water vapor and CO_2 from the air entering the respirometry system. This allows rates of CO_2 emission and water vapor loss to be measured from a zero baseline. Even if only CO_2 is of interest, it's not practical to remove CO_2 without removing water vapor, whether the scrubber is chemical or not. If a relatively high humidity is required without CO_2, it's easiest to remove both CO_2 and water vapor from the air and then add water vapor back in a controlled fashion (see Chapter 4).

The chemical costs of prescrubbing large volumes of air can add up, and if you require this capability you might want to invest in a regenerating scrubber, also known as a purge gas generator. Balston is a well-established maker, but others exist. These devices require a high-pressure compressed air line and a grid power source but will tirelessly and indefinitely produce a nearly dry and CO_2-free air stream of about 1–2 L min^{-1}. The tiny remaining amount of water vapor and CO_2 can be removed with a chemical scrubber which will last for months of use. One caveat: The O_2 output level of these devices may fluctuate slightly when they swap columns for regeneration, so it is good practice to direct the regenerating scrubber's output to a 10-L or larger carboy that acts as a buffer (open to the room via a short section of tubing so that it does not build up pressure; see Figure 4.2). Scrubbed air can then be pumped from the carboy as required.

Another approach to prescrubbing air for water vapor removal is to use condensation. This approach is described further in the section "Thermal scrubbers for water vapor."

CHEMICAL SCRUBBERS

Chemical scrubbers are something of a bugbear, but they are one of the traditional accoutrements of respirometry. Especially before the advent of accurate CO_2 analyzers and water vapor analyzers, it was impractical to compensate mathematically for the various interactions among O_2, CO_2, and water vapor. Scrubbers gave everyone a simple and honest way of eliminating troublesome terms from equations.

Tl/dr: Chemical scrubbers remain useful, and you need to know about their various characteristics and idiosyncrasies. But if practical, you should seriously consider mathematical correction for water vapor dilution, a topic that is thoroughly covered in the section on "Mathematical scrubbing" in Chapter 8.

The physical housing of chemical desiccant columns is not critical. For large-volume drying, prefabricated columns made from acrylic and prefilled with Drierite are available, and these can be reused for other desiccants as well. Otherwise, a 20- to 30-cm length of 2- to 3-cm-diameter acrylic tubing with holed stoppers will make a useful general-purpose, high-capacity column for any scrubbing agent. The inlet and outlet should be filtered with cotton or glass wool, and different chemical scrubbing agents (especially Drierite and Ascarite) should also be separated from each other by a thin layer of cotton- or glass wool. For subsampling in systems where response speed is important, especially for multiplexing between respirometry chambers, smaller scrubbers must be used, but they will tend to wear out quickly. Investigators may wish to consider nonchemical scrubbers for such applications (see the section on "Thermal scrubbers for water vapor," and especially Chapter 8's discussion of mathematical techniques for water vapor dilution correction).

Chemical scrubbers for water vapor

Several water vapor scrubber chemicals exist. Some can be quite easily restored to their dry state; others cannot.

Drierite

Drierite is the trade name for granular calcium sulfate manufactured by the W. A. Hammond Drierite Company of Xenia, Ohio. It is an excellent general-purpose desiccant with some caveats, which I'll address below. Typical respirometry applications use 8-mesh granule size, or for low flow rates and small columns the 10- to 20-mesh size. Regular Drierite is white and gives no indication of its hydration state; indicating Drierite, which is about 25 percent more expensive, turns from blue to pink when hydrated, thanks to an admixture of cobalt chloride. When fresh, Drierite can remove all but about $5 \, \mu g \, L^{-1}$ of water from the air, making it very effective for eliminating water vapor terms from equations. Moreover, it can be regenerated by spreading it in a thin layer and heating it for an hour at 200–220°C. It's important to note that lower temperatures won't break the bonds that hold the hydration in place, but significantly higher temperatures will alter the crystal structure of the calcium sulfate, ruining its hygroscopic proclivities. The regenerated Drierite should be placed in its original glass bottles while still warm. In practice, it generates some powder during regeneration and thus should be sieved first. Regeneration also causes the cobalt chloride to migrate into the Drierite, making the brisk blue-to-pink change more like a dull, faint blueish-gray blurring into a dull, faint gray. Amending the regenerated Drierite with some fresh indicating Drierite helps solve this problem.

Nice though it is, Drierite is not perfect. It tends to adsorb CO_2 reversibly, leading to maddening sluggishness in CO_2 analyzers whose sample streams are dehydrated with Drierite. A large column of Drierite can cause the baselining of a flow-through respirometry system to morph, where CO_2 is concerned, from a fairly low-stress, perfunctory event to a frustrating saga requiring 10 min or more. This has a secondary but still potentially significant side-effect even when only O_2 is being measured because if significant amounts of CO_2 are being slowly released during a baselining event *and if CO_2 is not being absorbed*, the additional CO_2 will dilute the O_2, causing it, too, to creep slowly to a stable value over the course of several minutes. This is a much less marked effect than on CO_2 but can nevertheless be annoying. The bottom line is that unless you will be absorbing CO_2, Drierite cannot be wholeheartedly recommended. If you are measuring CO_2 itself, then Drierite should be avoided.

Silica gel

Silica gel is an interesting substance made by creating a colloid of silicic acid, which forms when sulfuric acid is added to sodium silicate. The colloid is washed, dried, and formed into glassy granules. It has a huge surface area, with pores about 3–60 Å in size, and it works entirely by adsorption and by capillary action (within tiny pores, the saturation vapor pressure of water is greatly reduced). In its desiccant state, silica gel is amended with cobalt chloride so that it changes from a deep blue when dry to pink when its internal moisture content rises above about 8–10 percent by weight. Although silica gel can absorb a lot of water (up to 40 percent of its weight), it's not very effective at reducing water vapor to the levels possible with a desiccant that chemically reacts with water. In addition, its efficiency falls rapidly at temperatures $> 25°C$, and it functions more efficiently at high humidities, so that the drying ability of a silica gel column declines rapidly while the gas becomes drier as it migrates further into the column. All told, silica gel is an underwhelming desiccant, although it's certainly better than nothing.

Different grades and manufacturers of silica gel recommend different drying regimes, mostly based on pore size. A regeneration temperature of 120°C for at least 2 h is common. Fully saturated silica gel may require up to 6 h to dry completely. The blue color will return quickly but does not indicate that the regeneration process is complete. This is because fully dried silica gel has a water content of approximately 2 percent w/w; the color change takes place when the water content drops below about 8 percent by weight, so it still has a way to go, even though it looks fully restored. Microwave heating can be used but tends to be unpredictable and may damage the colloid. Blackish or brownish discoloration indicates overheating. As with Drierite, the silica gel should be bottled when hot. Regeneration does not diminish its indicating ability.

Because of its porous and highly adsorbent nature, silica gel should be scrupulously avoided where CO_2 is being measured, and it should be used with caution in O_2-based respirometry unless CO_2 is absorbed.

Magnesium perchlorate

At this point you may be wondering what chemical dryers *are* compatible with CO_2. Only one can be recommended, but with some strong caveats.

Magnesium perchlorate is an effective and rather frightening desiccant. As described by Willard and Smith (1922), it is as effective a desiccant as phosphorus pentoxide, the gold-standard desiccant and an even more frightening chemical. It can be regenerated as described by Willard and Smith, but the process is not recommended for routine laboratory use. However, it will not react with CO_2 and is pretty much the only desiccant to use if you want to dry air for CO_2 analysis using a chemical scrubber. Its extreme efficiency as a desiccant yields beautifully stable, low-humidity baselines for demanding water vapor analysis applications such as quantifying real-time water loss from a single *Drosophila melanogaster* (see Figure 16.2).

The main disadvantage of magnesium perchlorate is that it is an extremely powerful oxidizer and thus reacts vigorously with many common materials, including your skin and the lining of your respiratory tract. If you add water to it directly, it will hiss like a snake. During regular use, a magnesium perchlorate column will slowly start to deliquesce, eventually turning, if you are oblivious to its state, into a toxic creeping sludge that will set about invading and damaging your gas analyzers and, should you offer it the opportunity, yourself. Mitigating this tendency, magnesium perchlorate has a high capacity for water vapor uptake, binding up to 60 percent or more of its initial mass. It should be watched carefully and disposed of as soon as it starts to deliquesce. Note: it should be disposed of in accordance with local regulations; don't throw it in the trash because the perchlorate ion is regarded as a significant and persistent environmental pollutant. I am specifically not suggesting that you take up amateur (or professional) rocketry as a way of having fun with your used-up desiccant.

When selecting magnesium perchlorate, specify a granular type such as Anhydrone. Powders do not allow the efficient transfer of gases through the column. A perlite-based indicating magnesium perchlorate desiccant has been described (Wu and He, 1994), but to the best of my knowledge it is not yet commercially available.

Chemical scrubbers for carbon dioxide

There is no easy way to remove just CO_2 from the air. Water vapor will invariably be affected as well, so, generally speaking, CO_2 and water vapor scrubbers are usually used in conjunction (but the separate chemicals should never be mixed together). Usually water vapor is removed first, then CO_2. All CO_2 absorbents release water vapor as a byproduct, so it is good practice to include a final desiccant column after the CO_2 absorbent. Note that no CO_2 absorbent can be regenerated.

Soda lime

Soda lime is a mongrel compound created by slaking quick lime (calcium oxide) with a concentrated sodium hydroxide solution. The result is formed into pellets and consists of about 75 percent calcium hydroxide and 4 percent sodium and potassium hydroxide, with the balance being bound water. In some products, ethyl violet is added as an indicator. As sodium hydroxide is consumed, the pH of the soda lime falls, and when it drops below ethyl violet's critical pH of 10.3, a sudden shift to violet occurs, signifying the end of the product's useful life. Other makers of indicating soda lime use a brown-to-white shift to signal exhaustion. Soda lime is a fairly efficient CO_2 scrubber and is easy to handle with basic precautions. It is not as efficient as Ascarite, but it is much cheaper and does not deliquesce as easily. If a very stable zero-CO_2 air stream is required, it can be pre-treated with soda lime and the small remaining amount of CO_2 mopped up with Ascarite.

Ascarite

Ascarite used to be made of asbestos coated with sodium hydroxide and was less than pleasant to breathe in by accident. Even now, with the asbestos replaced with silica (vermiculite), in celebration of which the product was renamed Ascarite II, it should be treated with caution. Gloves and a mask should be used if you are not comfortable with touching or breathing caustic soda-coated silica dust. Thanks to its very large surface area, Ascarite is an extremely effective CO_2 scrubber. Because the CO_2 it captures is transformed to sodium carbonate, which is white as opposed to Ascarite's tan color, Ascarite contains a built-in indicator.

Ascarite should *never* be exposed to moist air, which in practice means air passing through a depleted desiccating scrubber. If Ascarite is exposed to moist air, it will deliquesce and start to creep, causing general inconvenience and destruction. If it creeps into Drierite, interesting chemical reactions will occur. It will also react vigorously with anything made from aluminum. Alternatively, when exposed to moist and then to dry air, it may solidify and block all air flow. It should always be used with a preceding desiccant column, and the state of that column should be checked frequently and changed before the depleted zone reaches the Ascarite.

THERMAL SCRUBBERS FOR WATER VAPOR

Water vapor will spontaneously condense into liquid water at and below its dewpoint. This property can be used to make water vapor scrubbers of varying efficiency, depending on the exact dewpoint chosen. Obviously, the lower the dewpoint, the better. The problem, of course, is that if a dewpoint below 0°C is chosen (technically an ice point), the ice will accumulate until it blocks the dryer. Thus, most thermal scrubbers use a Peltier effect heat pump to drop the temperature of a condenser to just above freezing, at no less than 1°C, at which the water vapor pressure will be about 0.65 kPa, but more commonly 5°C, corresponding to a WVP of ~ 0.9 kPa. This corresponds to a relative humidity of about 25–35 percent at room temperature, which is not impressively low. However, the condensed water can be drained, and large volumes of air can be dried in this way, with the balance of the water vapor being removed by other means if necessary.

The thermal scrubber is best viewed as an ancillary method for drying air, and it suffers from an additional disadvantage that might or might not be problematic in your application, depending on the speed of response you require for CO_2 analysis. CO_2 is highly soluble in water, and some CO_2 in the air stream will be lost with the condensate. This is unlikely to be significant but should be borne in mind as a possible error source.

If you want to thermally dry the air below $0°C$, it must be viewed as a temporary measure, after which the ice within the cold tubing will need to be thawed and drained. Copper tubing coiled into a helix is traditionally used. You can reach usefully low temperatures with a refrigerating water bath; alternatively, you can use liquid air ($-190°C$), a dry ice and acetone slurry ($-78°C$), or an alcohol and water ice slurry. *Don't use liquid nitrogen*; the O_2 in the air stream will liquefy and create an explosion hazard. Be prepared, with any of these mixtures, for the inevitable moment when flow will rapidly decline and then stop. Monitoring postscrubber flow rate, even if only with a rotameter, is essential. Also, be certain to allow the air stream to come back to thermal equilibrium with ambient conditions before it is allowed to enter instrumentation.

SELECTIVE MEMBRANE SCRUBBERS FOR WATER VAPOR

An interesting and effective dryer can be made from a membrane that is effectively permeable only to water vapor. Such a membrane does, in fact, exist, originally developed by DuPont and sold under the name Nafion. Nafion is a copolymeric mixture of a fluorocarbon, tetrafluoroethylene (Teflon), and perfluoro-3,6-dioxa-4-methyl-7-octene-sulfonic acid. The sulfonic acid groups create ionic channels through the membrane, along which water molecules can migrate very rapidly, following any transmembrane water vapor concentration gradients. Because this reaction is totally specific to water vapor and the fluorocarbon matrix is chemically inert, the other gases passing by the membrane (nitrogen, oxygen, and carbon dioxide) are unaffected.

Thus, you must pass your sample gas through Nafion tubing and attempt to have as low a water vapor pressure as possible on the outside of the tubing, into which the moisture from your air sample will pass. There are several ways to accomplish this. The easiest way is simply to surround the tubing with a desiccant, and in fact some rather clueless individuals and companies do just this. Nothing is gained from this exercise relative to using the desiccant directly, other than a slight increase in speed of response because of the tubing's lower internal volume. Set against this is the fact that the desiccant near the tubing will be rapidly exhausted, and there is no process other than agonizingly slow diffusion operating within the desiccant column to help spread the load. The result is good performance for a short time, followed by a rapid and prolonged reduction in drying efficiency.

Alternative approaches usually involve creating a concentric set of tubing, with the Nafion tubing running inside another piece of tubing, so that you can control and replenish its external environment easily. You can flush the outer lumen with dried air (from an autoregenerating dryer as described in the section "Prescrubbing air," for example), or with a dry tank gas at a low flow rate. Another possibility is to create a very low pressure in the outer lumen, typically by using a pump. One way to do this is to attach the outer lumen to an in-house vacuum supply if you are fortunate enough to have one. Some commercially available dryers use a reflux technique, in which sample gas is drawn through the Nafion tubing, into a gas analyzer (or analyzers), and then out through a restriction valve. The restriction valve is connected to the outer lumen of the tubing and is adjusted so that an internal pump pulls the air through the outer lumen at reduced pressure countercurrent to the sample flow. If properly implemented, this reflux technique can dry air that is saturated with water vapor at 20°C to a respectable dewpoint of –10°C, indefinitely. The lack of any consumable chemicals makes this drying technique quite attractive, quite apart from its low internal volume and resulting rapid response times. If a still lower dewpoint is required, a chemical scrubber such as a magnesium perchlorate column can polish off the remaining water vapor and will last for a long time.

Note that Nafion tubing exposed to ambient air will slowly develop a beige, then brown, and finally black discoloration. According to the manufacturer this does not affect its performance.

The limitations to the drying ability of Nafion tubing are in the volume of air that can be feasibly dried per unit time and its drying efficiency. To maintain a high surface area to volume ratio, Nafion tubing is usually very small in diameter, about 1 mm. It can be connected in parallel for higher flow rates, but this rapidly becomes uneconomical. It is therefore only suitable for drying small flow rates, up to about 200 mL min^{-1}, unless very expensive parallel implementations are used. This is adequate for subsampling larger flow rates or for drying the primary flow in smaller systems. In terms of drying efficiency, the sulfonic acid groups will retain some water vapor at all times, and this sets a lower limit to the minimum water vapor pressure the tubing can attain, even if the water vapor pressure on the nonsample side of the tubing is zero. This effect is highly temperature dependent. The minimum dewpoint of a Nafion-dried air stream is about –40°C, corresponding to 75 ppm water vapor at sea level. This is at 20°C. As the temperature of the tubing increases, the sulfonic acid groups embrace more and more water molecules, elevating the attainable dewpoint by about 1°C for each degree increase in ambient temperature.

TUBING AND TUBING CONNECTORS

The informed selection of tubing can have a huge impact on the usability and data integrity of a flow-through respirometry system. If you have ever watched the displayed water vapor content of an air stream decline sluggishly over a period of several minutes in response to a flow of dry nitrogen and then noticed that part of the plumbing was made of a softened elastomer such as Tygon, replaced that tubing with something more suitable, and were astonished at the massive decrease in response times that resulted, you won't need convincing. After considering tubing diameter, I present a list of tubing that is available in commonly used sizes and is often used (rightly or wrongly) in respirometry systems.

Tubing diameter

Tubing diameter is an important consideration. To minimize response times, you will want to keep tubing diameter as small as feasible. However, bear in mind that *flow resistance is inversely proportional to the fourth power of tubing radius*. Thus, reducing tubing diameter by half *increases flow resistance by a factor of 16*. The judicious use of italics emphasizes that for respirometry systems with high flow rates, using unnecessarily small-diameter tubing will cause large pressure offsets and corresponding extra work for the flow-delivery system. As a very general suggestion, for short tubing runs up to a meter or two in length you can use the following equation:

$$MF = 15D^4$$

(19.1)

where MF is the maximum recommended flow rate in milliliters per minute and D is the tubing's inner diameter in millimeters. Thus, 3-mm or 1/8-inch i.d. tubing can be used for short runs (a meter or two) up to about 1200 mL min^{-1}, whereas for 1.5-mm or 1/16-inch i.d. tubing, the corresponding recommendation is to keep flow rates < 75 mL min^{-1}. Six-millimeter i.d. or 1/4-inch tubing can be used at up to about 20 L min^{-1}. This is only a very general guide and invites exceptions based on your application. As a rule, however, you should keep the flow rate through gas analyzers below 200 mL min^{-1} to minimize pressure effects, unless you know that the analyzer in question can take higher flow rates *and* it is calibrated at the flow rate you will use.

You can estimate the approximate flow rate you will need to use in an experiment from equations derived in the section "Flow rate: a gas analysis perspective" of Chapter 8.

Tubing materials

Stainless steel is, of course, the best tubing in terms of performance and the worst in terms of usability. I mention it here only to acknowledge this fact. If you intend to use it, do so only for fixed setups that will not need to be changed frequently, and be prepared to spend a great deal of time tracking down leaks discovered during pressure testing.

Tygon and other similar soft to quasi-rigid tubing is certainly the most user friendly. Composed of clear PVC polymer with di(2-ethylhexyl)phthalate (DEHP) added as a plasticizer, it slips easily over tubing connector barbs, can sustain tight bends without collapsing, and is generally a joy to use. It also has very poor performance with water vapor, resulting in intolerably slow response times for water vapor analysis and indifferent performance with CO_2. It can be used with respirometry systems in which water vapor is scrubbed before analysis. If CO_2 is scrubbed as well, there is probably little reason not to use it, although the phthalate content is something of a concern because it is an estrogen mimic. Phthalate-free formulations produced in response to the Reduction of Hazardous Substances (RoHS) initiative in the European Union can be obtained.

Silicone tubing is wonderful to use but should be avoided because of its high gas permeability. You sometimes see it used in commercial respirometry systems in the biomedical field; the manufacturer is blissfully unaware that its use is a proclamation of incompetence.

PharMed tubing is advertised as having extremely low gas permeability, but its water vapor performance is indifferent and its price is remarkably high. It is composed of a polypropylene polymer with a mineral oil plasticizer. It is probably a better choice than Tygon for most setups; however, Tygon is transparent, allowing dust, condensation, or escaped fruit flies (it happens) to be seen, so the choice is difficult.

Tygon inert tubing solves the poor water vapor and CO_2 performance of normal Tygon tubing by lining its interior with fluorinated ethylene propylene (FEP), a chemically inert, intensely hydrophobic resin. It is clear, flexible, and expensive. Because the FEP liner is not intrinsic to the outer tubing polymer, it tends to delaminate if inappropriate connectors are used. Thus, this tubing should only be used with single-barb connectors or compression fittings.

Teflon tubing is translucent and rigid. Its excellent water vapor and CO_2 performance make it a good choice for critical small-animal respirometry. However, it is very difficult to use, and constructing a leak-free system based on Teflon tubing is not a trivial task. It is also extremely expensive. It will need to be used with compression fittings and John Guest fittings (see the section "Tubing connectors"); it is too rigid to allow the easy use of barb fittings.

Bev-a-Line tubing is a good all-round choice, compromising between low permeability and mechanical rigidity. It has excellent water vapor and CO_2 performance, thanks to its integrated high-density polyethylene liner. Its translucent ethyl vinyl acetate outer layer has no plasticizers but is still fairly flexible. Because its inner layer and outer shell are coextruded, they do not separate when the tubing is applied to a barb connector, though it should be noted that after being applied, the only way for anyone other than Arnold Schwarzenegger to remove it is by carefully cutting it off with a razor blade. To make connector insertion easier, heating Bev-a-Line to about 60°C makes it noticeably easier to work with. Too high a temperature will make it too soft (it will become nearly transparent at that point); use your judgment.

Tubing connectors

Tubing connectors (also known as tubing fittings) can make a surprisingly large difference to your experiment. Many respirometry systems, even in well-established labs that should know better, are cobbled together using hardware-store connectors intended for gas line connections that are oversized and clumsily mismatched for respirometry. There are better ways to occupy your time than trying to force 3-mm i.d. tubing onto a connector intended for 6-mm i.d. tubing. For very high flow rates you can simply use

PVC tubing connectors available from hardware stores; for other applications the following is a simple guide to four classes of connectors that together are adequate for at least 95 percent of small-animal setups. (You will find tubing connector suppliers in Appendix 1.)

First, *bulkhead connectors* are for mounting into the panel or "bulkhead" of an instrument or fixed setup, including the wall of a respirometry chamber. I recommend those made by the Clippard Instrument Laboratory. These consist of an internally and externally threaded metal body that passes through a chamber wall and is secured in place with a nut. The internal thread accepts any of a collection of threaded connectors (barb and compression) suitable for a wide range of tubing. The 3-mm barb connectors are probably the most useful.

Second, *interchangeable tubing-size connectors* are invaluable. They consist of a polyethylene barb fitting attached to either a male or a female tapered, push-fit body. I recommend Bel-Art "two-piece quick disconnects." The beauty of these connectors is that they are available with barb fittings for tubing ranging from 2 to 12 mm i.d. with the same push-fit bodies on all. Thus, adapting one tubing size to another is fast and painless and there is no need to keep a vast stock of mutually incompatible connectors at hand for each size of tubing you use.

Third, *John Guest connectors* are intended for rigid or semi-rigid tubing such as Bev-a-Line or Teflon tubing. The outside of the tubing is very lightly lubricated with vacuum grease (an imperceptibly thin layer) and inserted into the fitting, where it is sealed by an O-ring and held in place by a toothed, stainless-steel ring. Tubing can easily be released by pushing on a collar that disengages the toothed ring. They are available in straight-through, T, elbow, and bulkhead styles in a wide variety of sizes for fitting different tubing diameters. John Guest connectors have excellent sealing properties and greatly simplify the use of semi-rigid and rigid tubing such as Bev-a-Line, copper, or stainless steel.

Fourth, *Luer connectors* use the familiar small, tapered male and female fittings found on hypodermic syringes and their needles. Originally invented by a German instrument maker living in Paris, H. Wolfing Luer, they were an improvement on the earlier injection technology of Sir Christopher Wren, the architect, who favored goose quills, and were further improved in the early twentieth century by Farleigh Dickinson of the Becton-Dickinson company in the United States. The Luer fitting, being tapered, is intrinsically leakproof if maintained in good condition and tightened with an assertive twist. Intended for flexible or semi-rigid tubing in 3 or 1.5 mm i.d. sizes only, their other end has a barb fitting. These fittings are extremely useful for any respirometry system, or sections of a respirometry system, where the flow rate is less than a couple of liters per minute (higher if higher driving pressures are tolerable). Allied fittings (T, elbow, reducers, and so on) are available from the same manufacturer. These fittings are probably all that you will need for most of your low- and medium-flow-rate work.

Do not give way to the temptation to insert a barb fitting into a female Luer connector. This is usually done by people who insert the male side (not the barbed side) of a male Luer fitting into tubing, then while mouth-breathing, and after staring at it for a while, and after dragging their knuckles on the floor and trying to figure out what to do next, insert its barbed end into a female Luer connector. This causes leaks. I am mentioning this because even though it is a silly practice, it is widespread among the technically challenged.

A gender convention that is extremely useful: Always apply connectors to tubing while keeping the direction of flow in mind. Conventionally, this means using male connectors where the flow leaves the tubing and female connectors where the flow enters. If you feel that this encourages cisgender, heteronormative sexual stereotyping, and the perpetuation of the patriarchal hegemony, you are welcome to invert the convention. The important thing is to be consistent because this greatly facilitates interchanging sections of systems. For example, scrubber columns should always be plumbed so that flow enters the more easily removed end, so that exhausted chemicals can be quickly replaced; filters, many designs of respirometry chambers, and almost all gas analyzers have a well-defined input and output polarity. Maintaining a connector gender convention makes correct connections (such as plumbing a drying column the right way around) and interchanging sections of different setups far easier and more foolproof, or at least less foolprone.

20

Acquiring Useful Tools and Skills

There is a lot of material in this book. My intention in this brief chapter is not to over-whelm the reader, but to point a way forward for the small minority of researchers who might think, "Hey! Maybe there's a way to automate this process," or "I bet I can approach this research challenge in a different way, but I haven't seen any commercially available equipment that allows me to do that." When the first edition of this book was published in 2008, electronics, microcomputers, 3D printing, and computer-assisted design were all the domain of extreme nerds (*c'est moi*), well-funded, high-tech labs at places like the Massachusetts Institute of Technology (MIT), and large companies. Thanks to the maker movement these technologies are becoming democratized. All of them can find productive use in any innovative research laboratory at reasonable cost. Moreover, a burgeoning open source movement has synergized with the maker move-ment to promote the free sharing of equipment ideas across research laboratories. Following is an idiosyncratic list of technologies you might find useful.

Please understand that while I love your technical ambition and desire to learn, I am not available for technical help with your projects based on what you read here. This isn't from a false sense of grandeur—I simply don't have the time!

PROGRAMMING LANGUAGES

You might or might not already know how to program. If you do, excellent. If not, you should because it opens a wide vista of data analysis and data handling opportunities you might otherwise miss. Proximately, it will look good on your resume; ultimately it may make you a better, more creative researcher able to create and improve tools rather than passively use those that others have made. Here is a short list.

BASIC: Don't even think about it. It is a low-spreading, tangled tree that is easy to enter but frustrating to climb. I have written code in many dialects of BASIC, including Sinclair ZX81, Apple 2, GW BASIC, BBC Micro, QuickBasic, BASIC Pro 7, Visual Basic versions 2–6, and Visual Basic.net. I dearly wish I had never touched the damn language after about 1987 and respectfully urge you not to do so either.

C, C++, and C#: If you intend to write code for which speed is all-important, the various dialects of C allow you to program at the "bare metal" level, which is to say just a hop

Measuring Metabolic Rates: A Manual for Scientists. Second Edition. John R. B. Lighton, Oxford University Press (2019).
© John R. B. Lighton (2019). DOI: 10.1093/oso/9780198830399.001.0001

(technically, a compilation) away from the machine language that is your computer's or microprocessor's native tongue—but which is your worst nightmare of a programming language. The advantage of C is speed because it is compiled; the disadvantage is exposure to the ugly details of how the computer handles data at its most fundamental level. It's worth learning if you're willing to put in the effort and/or need to write fast code. It is widely used for programming microcontrollers. The ++ and # flavors are object-oriented, making them more in synch with current programming practices.

Python: This is an easy language for beginners to learn, yet object-oriented behind the scenes and powerful enough to lead you on to almost any level of expertise. It is free, open source, and supported by a huge if somewhat fractious on-line community. Its scientific and data handling capabilities are legendary and growing by the minute. Slimmed-down versions (MicroPython and CircuitPython) are available for microcontrollers. Unlike C which is compiled, Python is interpreted a line at a time and is thus slower. But, computation-intensive routines are usually confined to libraries that are written in C and are blindingly fast. See Appendix 3 for Python distribution package suggestions. It runs on practically all operating systems including tiny, cheap single board computers such as the Raspberry Pi line. With a bit of electronics expertise you can use it for data acquisition and instrument control.

STATISTICAL ANALYSIS PROGRAMS

R (www.r-project.org): An extraordinarily powerful, open source statistical programming and analysis environment featuring a steep learning curve. As with many open source programs, a large user community exists and help can be found for most questions on-line.

Jamovi (www.jamovi.org): A free statistics platform featuring an intuitive user interface driven by an R powerplant behind the scenes. Optionally, you can interface directly with the R engine and "roll your own" statistical tests.

Python: (See also the section "Programming languages.") Arguably it is even more powerful than R and beginning to eclipse it. Python has an extensive range of libraries that cover practically every imaginable scientific and technical need, including production of publication-quality graphics, advanced statistical functions (as well as basic ones), machine learning, artificial intelligence (AI)—the list goes on and on. A huge on-line community is available.

SPSS, MATLAB, and to a lesser extent **LabView**: Good but proprietary and remarkably expensive packages unless you are lucky enough to have a site license. In my opinion, large-scale crowdsourcing by the open source community is rapidly outstripping the usefulness of these older packages.

It's always worth checking in at the OpenScience Project (www.openscience.org) for the latest open source software.

MICROCONTROLLERS

Think of a microcontroller as a stripped-down, tiny computer with limited memory, relatively low speed, and no operating system. It will obediently run any code you upload into it, generally using a dialect of C, though other languages are available. Thanks to the Arduino movement and its allies, microcontrollers have become versatile and approachable enough for the motivated researcher to use. Potential applications for these devices in most laboratories are legion; Google "Arduino open source laboratory" for some ideas. Bear in mind that any equipment you create will not be as rugged, well constructed, well documented, or FCC/CE certified or supported as commercially purchased equipment. But if no commercial options are available for your intended application, or you simply cannot afford them but need to get seed data for your next grant, microcontrollers are your friend.

> **Arduino** (www.arcuino.cc): Practically unknown when the first edition of this book was published, the Arduino lit a raging fire of innovation in the maker community. It is easy to learn, widely used in laboratory automation projects, and surprisingly powerful. Its on-line community is enormous. Arduinos are available in many flavors ranging widely in power and cost from a variety of manufacturers including the original Italian group that designed the first version. All of them can read voltages with moderate resolution and can control instruments, if you know what you are doing. Capabilities can be expanded via breadboarding or plug-in boards ("shields"). Programming these devices is generally done in the C language which is part of their free development environment. A wide variety of kits, videos, and books are available to get you started.

Listing all current microcontrollers and their associated environments is an exercise in futility, given the exponential increase in the number of products (and their power) available. Some creative searching on the Internet will be time well spent.

> **Raspberry Pi** (www.raspberrypi.org): Not really a microcontroller but a single-board computer with respectable speed and memory, running the free Linux operating system or, optionally, the Microsoft .NET framework. A couple of versions exist, with, again, a huge on-line community. Plug-in boards ("hats") or breadboarding give you access to voltage measurement and instrument control but require moderate electronic expertise. Programming is via the included Python distribution (Linux) or C# (.NET).

RESEARCHER-FRIENDLY ELECTRONICS HUBS

I pondered what to call electronics suppliers that are specifically friendly to the small-scale, hobbyist/speculative lab user. The electronics suppliers listed in Appendix 1 will supply almost any imaginable component, but you need to know *precisely* what you need. They are totally uninterested in what you are doing or in helping you to achieve anything. In contrast, a growing number of companies cater to individuals, place a premium on helpfulness, and have excellent on-line forums and helpful communities. They also trawl the ocean of worldwide electronics for the latest sensors, effectors, curious devices, and so on that might appeal to their customers. Many modern components are almost impossible to solder by hand, and these companies provide breakout boards that allow you to evaluate or use them.

SparkFun (www.sparkfun.com) was founded by Nathan Seidle in 2003 and sells a wide variety of kits, components, and breakout boards. It has lots of video tutorials and a good on-line community. Its designs are open source.

Adafruit Industries (www.adafruit.com) was founded by MIT graduate Limor Fried in 2005 (huge fanboi here!) and now dominates the maker-friendly electronics field. It sells an enormous variety of kits, components, sensors, breakout boards, and unusual items; most products have accompanying tutorials, including videos, that demonstrate step-by-step how to use them. It has a very active and inclusive on-line community. Its designs, like SparkFun's, are open source.

Polulu Robotics & Electronics (www.pololu.com) was founded by MIT graduates Jan and Candace Malášek in 2001. It sells a wide variety of sensors and effectors; as its name implies it specializes in robotics, so it is worth contacting for your mechatronic as well as general electronic needs. Not open source but very helpful and cooperative, Polulu also offers laser cutting services.

ELECTRONICS REFERENCES

The sources listed above will get you started, but you will soon require some electronics background to turn your grand ideas into reality. The good news is you do *not* require an electrical engineering degree—just an inquiring mind, willingness to learn, and the fortitude to endure and learn from your inevitable frustrating mistakes. In my opinion there is only one "how to" book in this area that is worth owning: Horowitz and Hill (2015, but earlier editions are OK). This is the standard bible of the field.

PRINTED CIRCUIT DESIGN AND MANUFACTURE

So, you have spent a few months learning electronics with the aid of water wings and training wheels. Now is the time for you to run with the big dogs, capture your own circuit diagrams, and design your own printed circuit boards. There are many options available; those below are recommended for you.

Eagle (https://www.autodesk.com/products/eagle/free-download): This is the *de rigueur* maker schematic capture and computer-assisted design and manufacture (CAD/CAM) package. It's free for personal or academic use in a stripped-down but serviceable form. The huge on-line community and many instructional videos on YouTube make it a logical choice for the beginner; its full versions are capable of very advanced designs. Open source hardware designs from SparkFun and Adafruit are all Eagle-based.

KiCad EDA (http://kicad-pcb.org): This is an open source, entirely free CAD/CAM package that runs on OS X, Windows, or Linux. I have not used it, but it appears to have powerful features and an engaged on-line community.

Oshpark (https://oshpark.com): This company's on-line portal will take a board file produced by either of the above packages and turn it into a splendid purple circuit board at a reasonable price.

It is beyond the scope of this very brief introduction to deal with fabrication techniques such as soldering, and especially the use of surface mount technology (SMT) components. Many modern components are only available in SMT form, and some of these simply cannot be hand-soldered. If you need to use them don't lose hope: search on-line for how-to guides or breakout boards.

3D DESIGN AND PRINTING

3D design and printing technology was prohibitively complex and expensive when the first edition of this book was published. Now it's routine. You can use free on-line software to create designs and print them using a service provider or a local 3D printer. Need a holder for a respirometry chamber, a clip to keep tubing tidy, a mount for that awkwardly sized object that keeps capsizing, an enclosure for your circuit board, and so on? It may take only a few minutes to design and an hour or two to print.

Tinkercad (www.tinkercad.com): This free, intuitive on-line service allows you to design basic 3D objects quickly in your web browser (Chrome is best for this). It has a large on-line community. You can download your CAD files for local 3D printing or submit them to on-line 3D printing service providers (Google them). Many instructional videos are available on YouTube.

Blender (www.blender.org): A free, open source 3D design package that is mostly designed for animation but can also be used for 3D design and printing. I have not tried it.

SketchUp (www.sketchup.com): Available in a free, browser-based and a "Pro" paid version. I find Tinkercad much more intuitive for simple designs but YMMV.

Fusion 360 (https://www.autodesk.com/products/fusion-360): This full-featured CAD package is cloud-based and free for users from educational institutions; others pay a modest monthly fee. If it can't be done in Fusion 360 (unlikely) you will need to move to SolidWorks with its astronomical purchase, maintenance, and support fees.

Thingiverse (www.thingiverse.com): A huge collection of 3D files for your printing pleasure. Pro tip: Search for "scientific" and be amazed at the variety of open source 3D designs for syringe pumps, pipettes, holder arms, Nunc trays, centrifuges(!), spectrographs—the list goes on. Many of these designs synergize with microcontrollers such as the Arduino to implement complex functions.

Joshua Pierce is a leader in the 3D scientific open source movement and has written a book about it—Open-Source Lab, published by Elsevier (2014). Drop in at www.appropedia.org/Open-source_Lab for the latest developments.

I will not list specific makes of 3D printers because they are in an exponentially increasing rate of flux—fortunately in the direction of improved features and lower prices. You can pick up a serviceable little printer for about US$300; even industrial-quality 3D printers with a large build volume and a vast array of compatible filaments have fallen below US$5000, an excellent investment if your lab can afford it.

LASER CUTTING

For quick fabrication of 2D material such as acrylic, cardboard, or thin plywood, laser cutters are an invaluable tool. You can create your design using any of the above CAD programs, or within simple 2D design programs such as Inkscape (free, open source), Adobe Illustrator, or CorelDRAW (which, oddly, is widely supported by laser cutter manufacturers). Need a quick test-tube rack, Eppendorf vial holder, tubing arranger, air-mixing baffle, or whatever? From starting the design to completing the cut, you can have it in your hands within half an hour or so. Among the more popular laser cutters are:

Full Spectrum Laser (www.fslaser.com): Excellent cost:performance ratio.

Epilog Laser (www.epiloglaser.com): High quality; many models available.

Glowforge (www.glowforge.com): Low-cost, easy-to-use cutters suitable for light work.

ENGAGE!

I hope you find this summary of lab-appropriate tools and technologies useful. I plan to keep it updated on this book's companion web site, www.respirometry.org. Please contact me if you have suggestions to include in the web site.

Long may you measure the fire of life! Success in your mission! Make it so! Engage!

APPENDIX 1
SUPPLIERS OF METABOLIC MEASUREMENT
SYSTEMS AND ALLIED APPARATUS

This Appendix is a brief, necessarily idiosyncratic list of useful equipment and suppliers relevant to research-grade respirometry and allied activities. I have concentrated on suppliers of which I have personal knowledge. A more complete and up-to-date list may be found at www.respirometry.org, where suggestions from readers for further items are welcome.

I will not give specific recommendations for suppliers of metabolic measurement systems and gas analyzers. To obtain information to guide your decisions, the Internet is a powerful tool, provided you maintain a skeptical attitude. Even more powerful is the recommendation of respected colleagues. Here are specific items to watch for:

- Do actual users of these devices recommend the firm in question? Often it is worth going behind the scenes and asking the technicians who use the system, rather than the Principal Investigators who operate at a remove and do not want the wisdom of their purchase decisions questioned.

- Most devices and systems are complex and will have issues from time to time. Does the firm provide prompt and helpful in-warranty support? For how long? What out-of-warranty support is available? Are costs reasonable or punitive?

- Does the firm make promises prior to purchase that it honors, or is there a history of delays, excuses, and broken promises?

- Are the firm's specifications free of exaggerations? (See "Calibration of CO_2 Analyzers" in Chapter 16 for some guidance on realistic specifications; beware of the Red Queen effect whereby firms, especially in the biomedical field, compete to exaggerate accuracy by orders of magnitude.)

- Does the firm provide prompt and expert technical support *and* advice on analytical questions regarding their systems? Maintaining a staff with this level of expertise is expensive. Most firms prefer not to make that investment, but it is a necessary resource for researchers competing for scarce grant funding.

Beware of the Stockholm syndrome. Quite commonly you will encounter researchers who use a system they despise but praise it to people who inquire about its performance. Their motive, usually unconscious, is twofold: to make it seem that their purchase decision was optimal, which reflects well on them; and to draw others, after they purchase it, into a comforting collective misery. We are a strange species.

ELECTRONIC DISTRIBUTORS AND SUPPLIERS

Allied (www.alliedelec.com): Distributes a wide variety of mainstream components.

Amazon (www.amazon.com): Yes, Amazon. Resells electronic components and kits from many suppliers. Wide range of dubiously authentic, obscurely sourced, but inexpensive breakout boards, most of which work and are OK for experimentation.

Digi-Key (www.digikey.com): Distributes probably the widest variety of components with an excellent selection of sensors. Exceptionally good on-line search capabilities and prompt delivery.

FindChips (www.findchips.com): Not strictly speaking a distributor, but rather a search engine that trawls all the major distributors to find a specific part for you. Useful for locating a distributor that actually has stock of a specific component, and for comparing prices.

Jameco (www.jameco.com): Carries an unusual selection of surplus and new items, many unique. Like most companies that emphasize their low prices, prices on some items may be somewhat higher than the competition's (see FindChips for a price comparison tool).

Mouser (www.mouser.com): Offers a wide range of components and hardware, including many niche development kits.

Newark (www.newark.com): Similar to Mouser. Emphasis is on the latest developments.

RS Components (www.rs-components.com): Distributes a wide variety of components, primarily UK based.

These are all professional/industrial distributers that offer excellent service but no application help or support. For a more human touch, see the firms recommended in Chapter 20.

TUBING, TUBING CONNECTORS, AND NEEDLE VALVES

Bel-Art Plastics (www.belart.com): Produces a wide range of plasticware, including excellent interchangeable barbed tubing connectors.

Clippard (www.clippard.com): Produces a wide variety of bulkhead fittings, barb and compression fittings, flow switches, and basic needle valves.

Cole-Parmer (www.coleparmer.com): Resells a wide range of tubing and connectors at fairly high prices.

McMaster-Carr (www.mcmaster.com): Resells a wide variety of tubing and connectors, including John Guest fittings; many large-bore options are also available.

Nordson Medical (www.nordsonmedical.com): Sells a wide range of mostly small-bore connectors and tubing. Very good website, reasonable prices, excellent quality.

Parker (www.parker.com/pneumatic): This industrial behemoth sells a wide variety of connectors, tubing, valves, and so on, mostly through distributors. Intricate website.

Swagelok/Nupro (www.swagelok.com): Sells a wide range of expensive, premium-quality needle valves and their close cousins, metering valves.

MASS FLOW METERS AND CONTROLLERS

Alicat Scientific (www.alicat.com): A wide variety of mass flow meters and controllers, operating on the differential pressure principle and boasting impressive accuracy, 100:1 turn-down ratio, and low differential pressure requirements.

Sensirion (www.sensirion.com): Accurate mass flow meters and controllers. The controllers require high differential pressures to operate correctly, so are more suited to use with tank gas than diaphragm pumps.

SEMI-PERMEABLE MEMBRANES

Membrana (www.membrana.com): Sells polymethylpentene tubing under the brand name Oxyplus. Useful for allowing the aerial diffusion of O_2 and CO_2, but not H_2O (see Chapter 5).

Perma Pure (www.permapure.com): Sells a wide variety of Nafion products, including single- and multi-lumen tubing, coaxial tubing, and so on. Can be used for drying or humidifying air; allows the aerial diffusion of H_2O but not O_2 and CO_2.

APPENDIX 2
THE CARE AND FEEDING OF CLARK
O$_2$ ELECTRODES

HOW CLARK ELECTRODES WORK

A Clark electrode consists of an anode typically made of chlorided silver and a cathode typically made from thin platinum wire. Only the very tip of the platinum cathode is exposed; the rest of the platinum wire forming the cathode is sealed in glass or epoxy. A membrane covers the tip of the platinum wire cathode, and beneath the membrane a thin layer of electrolyte allows electrical contact between the anode and the cathode.

The reaction occurring at the cathode is

$$O_2 + 4H^+ + 4e^- \rightarrow 2H_2O \tag{1}$$

and at the anode

$$4Ag + Cl^- \rightarrow 4AgCl + 4e^- \tag{2}$$

The molecular oxygen is consumed, with an accompanying flow of current through the cathode. This reaction takes place when the oxygen electrode is polarized over a range of voltages centered near 0.65 V. The current, I (amperes), that passes through an oxygen electrode is proportional, by Faraday's law, to the flux of oxygen, J (moles per second), arriving at the cathode and originating in the sample:

$$I = JnF \tag{3}$$

where n is the number of electrons (four) added to each oxygen molecule arriving at the cathode, thus reducing the oxygen, and F is the Faraday constant (96,485 coulombs mol^{-1}). Using this equation, you can also calculate how much oxygen a probe is consuming at a given current:

$$J = I / (nF) \tag{4}$$

Conditions at the cathode are so arranged that the partial pressure (or more strictly speaking, the fugacity) of oxygen is reduced to zero at its surface. The cathode can be said to be starved for oxygen, in that it consumes all the oxygen that arrives at its surface. This is an important point. And this is where the membrane (Clark's great innovation) comes in. By limiting the rate of oxygen diffusion from the external medium to the cathode, specifically to far less than the cathode's ability to consume oxygen, the membrane's characteristics determine the sensitivity, the speed of response, and even the linearity of the oxygen electrode.

Consider that a perfectly permeable membrane will overwhelm the cathode's ability to reduce the available O$_2$ at any but the lowest oxygen partial pressures, thus producing an unusably sensitive oxygen electrode. A perfectly impermeable membrane won't allow *any* oxygen to reach the cathode, leading to an unusable *insensitive* oxygen electrode. Between these extremes lies the territory of more-or-less usable oxygen electrode membranes, which allow enough, but not too much, oxygen to reach the cathode. Obviously, the bigger the cathode, the greater the amount of oxygen that can be reduced, and the greater the current (I) that the oxygen electrode will produce at a given oxygen partial pressure.

The more permeable the membrane,
the greater the speed with which oxygen reaches the cathode and is consumed
the more oxygen is consumed
the faster the response of the oxygen electrode
the more stirring is required to eliminate boundary layer effects.
The less permeable the membrane,
the slower the rate of oxygen consumption at the cathode
the less oxygen is consumed
the slower the response of the oxygen electrode
the less stirring is required.

Regarding stirring, if the partial pressure of oxygen reaches zero at the cathode, as we've just shown, then the oxygen partial pressure gradient must extend through the membrane and into the surrounding medium (or no oxygen flux to the cathode can occur), thus decreasing the oxygen partial pressure in the vicinity of the membrane. If this happens, the oxygen electrode is not measuring the partial pressure of oxygen in the medium; it is just measuring the partial pressure of oxygen in a thin boundary layer of the sample adjacent to the membrane into which the oxygen from the bulk of the sample diffuses. This is not terribly useful. To reduce or eliminate this boundary layer effect, stirring is required. Stirring is usually easy enough to organize, but there are times when it is not convenient. It can even be outright problematic with delicate organisms. But it's a fact of life with Clark electrodes. Deal with it or use fiber-optic O_2 probes or coulometry.

To quantify membrane permeability, first calculate the oxygen flux rate that is possible at the cathode:

$$J = \left(po_2\right)/\left[\left(Z_s/\left(D_s \times S_s\right)\right] + Z_m/\left(D_m \times S_m\right) + Z_e/\left(D_e \times S_e\right)\right. \tag{5}$$

where J is as before and po_2 is the partial pressure of oxygen in the sample (kilopascals). The three terms in the denominator are the resistances to diffusion from the sample all the way to the cathode, originating from the boundary layer at the membrane's surface (s), the membrane (m), and the electrolyte layer between the membrane and the cathode (e). Z refers to the thickness of the layer (in meters), D to the diffusion coefficient of oxygen in the layer (square meters per second), and S to the solubility of oxygen in the layer (moles per cubic meter per kilopascal). Obviously, if their sum total drops, J goes up, and J goes down if their sum total increases, regardless of po_2.

Let's interpret the significance of this equation. Equation 3 shows that the probe current is determined by J, and now we see that J is determined not only by the partial pressure of oxygen but also by the characteristics of the boundary layer, the membrane, and the electrolyte. Stability of the Clark electrode signal is the grail of aquatic respirometry. This means that to obtain stable readings, you need a stable electrolyte; you need a stable membrane; and you need a constant boundary layer, which translates to a constant rate of stirring or sample movement.

Table 1 Common Clark electrode membrane materials and their solubility and diffusion parameters, assuming a membrane thickness of 25.4 μm (0.001 inch).

Material	D_m (m² sec⁻¹) × 10¹¹	S_m (mol m⁻³ kPa⁻¹) × 10⁷	$Z_m/(D_m S_m)$
PTFE	2.54	0.106	0.943
PFA	2.69	0.093	1.02
FEP	1.73	0.088	1.67
Tefzel	1.34	0.017	112
PTME	343	0.040	1850

PTFE, Polytetrafluoroethylene; PFA, perfluoroalkoxy; FEP, fluorinated ethylene propylene; PTME, polytetramethylene.

Of the various resistances to diffusion listed above, by far the biggest is usually the membrane. This means that as the membrane becomes more permeable, the other resistances become more and more significant. So, for the best stability from the point of view of equation 5, you should use a low-permeability (usually meaning thicker) membrane and put up with the lower resulting J, meaning, in practice, slower response times. Table 1 shows some common membrane materials and their characteristics.

If the characteristics of the membrane change with time, or if the characteristics of the electrolyte change with time, you'll get a change in J, which means drift. Short-term changes mean noise. These can both come from other sources, too, as discussed in the section "Stability issues of Clark electrodes."

Thus, choose Teflon for fast response, fluorinated ethylene propylene (FEP) for medium response with less sensitivity to stirring rates, and Tefzel for slow response but very little sensitivity to stirring (put another way, compared to the resistance to diffusion possessed by the Tefzel, the boundary layer, $Z_s/(D_s S_s)$, is negligible). Incidentally, the thin polyethylene from cheap sandwich bags works rather well as an emergency Clark electrode membrane.

Let's sum this all up in terms of electrode current, which after all is what we're measuring. Neglecting diffusion resistance posed by the boundary layer and the electrolyte (translation: assuming the membrane outweighs them), we have

$$I = \left(nFA \times D_m \times S_m \times po_2 \right) / Z_m \tag{6}$$

where A is the area of the cathode in square meters and the other variables are as before. I can vary from fractions of a nanoamp (true microcathode electrodes), a couple of nanoamps (general micro- to semi-microcathode electrodes), through a dozen or so nanoamps (mesocathode electrodes) and up (conventional "button cathode" Clark electrodes). This is the master equation of an ideal oxygen electrode. Measured I will never be quite as great as calculated I, of course, because of the boundary layer and electrolyte terms. Try to keep all terms of equations 5 and 6 constant, except the one you want to measure—po_2.

Temperature effects are important as well. The current delivered by an oxygen electrode changes by about 3–4 percent per degree Celsius at constant oxygen partial pressure. This temperature coefficient is greatest for the least permeable membranes. Hence, you must be concerned not only with the stability of sensor output at constant temperature, but also with the stability of the temperature coefficient of oxygen solubility and diffusivity in the membrane. For this reason, most research work aims to keep the temperature of the system constant, rather than using a "fudge factor" temperature-correction term that may or may not be accurate for a given setup. Using black-box temperature compensation of an oxygen electrode is usually the mark of an amateur.

THE NECESSITY OF STIRRING

Having been exposed to a brief analysis of Clark electrode operation, you now understand the requirement for stirring and the requirement that the stirring rate remain roughly constant. These are unfortunate requirements (greatly mitigated if you are using a good fluorescence-based aquatic O_2 analyzer). Let's revisit stirring sensitivity and the speed of response tradeoff that comes with it. What about changing the cathode area rather than (or in addition to) membrane permeability? An alternative way to decrease the stirring requirement is to reduce the cathode diameter, resulting, in extreme cases, in the so-called microcathode electrode. Such electrodes consume very little oxygen and, as a result, produce very low currents. A couple of decades ago, scientists fell in love with microcathode electrodes because of their supposed freedom from boundary layer effects. Most scientists realize now that what they thought was love was actually infatuation. Why?

Because of their small cathode area, microcathode electrodes are plagued by fouling, poisoning, drift, noise, and other problems. Microcathode electrodes still require some stirring, contrary to what their manufacturers would have you believe. If you have to stir anyway, you might as well stir anyway, as Sam Goldwyn might have said. It's taken a long time for the scientific community to figure this out. A lot of people used (and still use) microcathode electrodes to monitor slowly changing oxygen partial pressures in constant volume respirometry applications, for example—a job for which they are less than ideally suited.

To reiterate, you can cut down on the stirring requirement by decreasing the permeability of the membrane. You can do this by specifying a less-permeable membrane material or a thicker membrane, at a cost in terms of speed of response. In all but a very few highly specialized applications this is not a problem. What you lose in terms of fast response, you gain in stability.

STABILITY ISSUES OF CLARK ELECTRODES

Finally, let's address stability, the overriding concern (and frustration) of oxygen electrode users. Two types of instability are observed during the operation of oxygen electrodes. First, there is often a gradual monotonic drift of sensitivity, usually in the sense that the current per unit oxygen partial pressure drops with time. Second, there are positive and negative fluctuations of sensitivity superimposed on the general drift. Some are of very short duration, of the order of seconds or minutes, and others are much longer, with time constants of the order of days.

The most important single factor causing drift is membrane tension. Because the membranes used for oxygen electrodes are normally thin and partially elastic, their thickness, and hence the resistance they offer to oxygen diffusion, can be substantially altered if they become stretched. This usually happens when they are being mounted. One of two deleterious effects may then occur. Either the membrane may gradually slip under the restraining means (usually an O-ring), to return to its original thickness, or it may suffer "cold flow" as a result of the constant stress, with a consequent steady change of thickness. These thickness changes often continue during a period of months after mounting and are observed as a gradual drift of output signal. True stability can only be achieved if membrane stretching is avoided during mounting. At the same time, you must not allow a thick electrolyte layer to remain between the membrane and the sensing cathode. This follows from the rule that the diffusion resistance of the electrolyte layer should be as small as possible relative to that of the membrane.

Here's a subtlety. I've just shown that the thickness of the electrolyte layer between the membrane and the cathode determines, in part, the sensitivity of an oxygen electrode. Hence, physical phenomena that directly influence the thickness of the electrolyte layer can give rise to instability. The membrane that encloses an oxygen electrode is semi-permeable in the sense employed in discussions of osmotic effects; that is, it allows the passage of gases, including water vapor, but blocks the passage of ions. When an oxygen electrode is immersed in pure water, therefore, water vapor enters the sensor, diluting the electrolyte. If the walls of the electrolyte chamber were completely rigid, the internal pressure would rise until the water vapor pressure in the electrolytic solution equaled the water vapor pressure in the water outside the sensor. Equilibrium would then be attained. In practice, however, the hydrostatic pressure inside the sensor at equilibrium, known as the osmotic pressure, can be as high as 104 kPa and thus cannot be restrained by the tension in the membrane. So movement of water into the sensor continues, with the membrane bulging outward to accommodate the increased volume of electrolyte. This entry of water into the sensor causes an increase in the electrolyte layer thickness, Z_e, and a consequent decrease in the output from the sensor.

To function properly, an oxygen electrode must be filled with an electrolyte having at least a minimum conductivity for current flow. This conductivity is decreased if the sensor dries out or if the electrolyte becomes diluted by sample water entering through a leak in the membrane sealing system. The symptom manifested by the sensor in this case is a gradual decrease of signal at constant oxygen concentration. To cure the problem, it is only necessary to renew the electrolyte and to ensure adequate sealing of the membrane to the sensor body.

Gravitational effects on the electrolyte layer can also occur. The membrane may sag to some extent if it supports the whole weight of the electrolyte. The effect is most significant if the electrolyte volume and the membrane area are large and if the membrane is poorly tensioned and supported. Gravitational effects can be minimized by clamping the sensor in a fixed orientation.

Vibrations of the sensor or turbulence in the sample sometimes cause the sensor to deliver a noisy signal. This problem is particularly acute if the membrane is poorly tensioned or if air bubbles are enclosed in the sensor. The explanation is thought to be that convection currents are caused in the electrolyte, thus transporting oxygen from the electrolyte reservoir to the cathode.

The ultimate determinant of performance is the electrical characteristics of the oxygen electrode itself. Oxygen electrodes are usually operated with a constant total applied voltage of about 0.65 V, shared between the anode–electrolyte interface, the cathode–electrolyte interface, and the electrolyte conductor. Ideally, the electrolyte solution should be an almost perfect conductor, and its share of the applied voltage should therefore be virtually zero. Also, ideally, the anode should be operated under equilibrium conditions so that its interfacial voltage is constant. Therefore, the cathode's interfacial voltage should remain constant, giving us a stable current and thus a stable po_2 reading, the Holy Grail of oxygen electrode users.

However, the electrolyte conductor may take an increasing share of the applied voltage with the passage of time, either because it becomes progressively depleted of salt due to the anode reaction or because it becomes gradually diluted with sample water due to faulty sealing of the membrane to the sensor body. To make things worse, the anode interface may take on a variable (rather than constant) fraction of the applied voltage, either because it becomes increasingly blocked by the deposition of nonporous insulating products or because electroactive vapors from the sample interfere with the anode reaction. In either case, the cathode voltage is also variable, usually in the sense that it decreases with time, causing a downward drift in sensor output.

The oxygen electrode reaction is also subject to interference by deposits on the cathode surface, either of electroplated metals (the anode metal is frequently plated onto the cathode during operation), organic substances adsorbed from the electrolyte, or products of side reactions at the cathode caused by interfering vapors from the sample. The smaller the cathode, in general, the worse this problem becomes. Again, a downward drift of sensor output at constant oxygen partial pressure is the usual consequence.

Ideally, the electrochemical reaction at the anode of an oxygen electrode should result in the creation of an insoluble porous solid phase that adheres firmly to the anode metal. In practice, the products generated at most of the commonly chosen anode metals, such as silver, lead, cadmium, thallium, or zinc, are soluble in the electrolyte to some extent because of the formation of complexes between the metal ions and the anions of the electrolyte. Diffusion of these ions and subsequent plating of the anode metal on the cathode is therefore a common phenomenon.

The plating is less rapid when the membrane is well tensioned and when the zone surrounding the cathode, through which diffusion of the metal ions occurs, is long and narrow. Also, the useful life of the sensor is longest, for any particular sample, for membranes of low permeability because the initial rate of formation of the metal ions is proportional to the membrane's permeability.

Sometimes the plating occurs over the whole area of the cathode, an effect that is most easily noticeable at gold cathodes because of the change of coloration. This is not too troublesome if the anode is silver, but it can lead to a decrease in cathode activity when the anode metal is more basic. The cathode should be polished in these circumstances to restore its activity. This is quite easy to do (see the section "Polishing the cathode").

In other circumstances, the plating occurs at the periphery of the cathode, as treelike dendritic formations, thus extending the sensing area and leading to an increase in displayed concentration at constant sample concentration. Again, the cathode must be polished at regular intervals if this becomes a problem (see the section "Polishing the cathode").

More serious is anode blocking. The continuous operating lifetime of an oxygen electrode is often limited by the blocking of its anode by nonporous reaction products. Anode blocking usually occurs when a particular quantity of anode product has been deposited on each unit area of the anode, and therefore the larger the area of the anode, the lower the permeability of the membrane, and the lower the concentration of oxygen to which the sensor is exposed, the longer the life of the sensor. After anode blocking has occurred, activity is easily restored by stripping the product off the anode with a suitable cleaning solution. A silver anode, for example, can be cleaned with a commercial detarnishing fluid or with an ammoniac solution. It will usually need rechloriding after this treatment (see the section "Rechloriding the anode").

When an oxygen electrode must be operated well above room temperature, extra care is necessary to avoid enclosing an air bubble in the sensor. The reason is that this bubble expands more than the displaced electrolyte would have done, and the additional volume might be created by "inflation" of the membrane. This, of course, leads to a diminished output from the sensor and sometimes to an irreversible stretching of the membrane.

The zero current of an oxygen electrode is the current produced by the sensor when exposed to a medium containing no oxygen. It is never zero, although the sensor should be selected so that it is negligible in comparison with the signals of interest in any particular study. Commercial sensors are available having zero currents in the range from 0.01 to 1 percent of the signal generated in air-saturated water. Hence, the stability of zero, or residual, currents is of little interest when measurements are made in well-aerated clean water, but it becomes crucially important when measurements are made in anaerobic media.

Even a cursory study of residual currents reveals that they are rarely constant in time. For example, a current spike followed by an exponentially decaying current tail is observed whenever a sensor is switched on after being exposed to air while off circuit. The time constant for this decay can be minutes or even hours, depending on the particular sensor and the state of cleanliness of the cathode. Beware, microcathode electrode users.

Because of the transient nature of the residual current, the practice of electronically adjusting the zero point of a sensor by subtraction of a constant current is of limited utility. The residual current is completely internal in origin and so becomes less important as the oxygen permeability of the membrane increases because the signal level increases in comparison with the constant error level.

Finally, let's look at some of the major compromises between the stability and speed of response of oxygen electrodes. I've mentioned membrane permeability in several places, in the context of oxygen electrode stability. This is because the oxygen must pass through the diffusion boundary layer and the electrolyte layers as well as the membrane, and these two additional resistances are likely to vary during operation; therefore, stability can be expected only when the membrane impedance exceeds those of the other two by a safe margin. This reasoning alone dictates a low-permeability membrane for an oxygen electrode, whenever stable, long-term monitoring of oxygen is important.

Another advantage accrues from the use of a low-permeability membrane: Only a low rate of stirring is required to reduce the impedance of the diffusion boundary layer to a negligible value in comparison with that of the membrane. Indeed, very high stirring rates are needed to supply oxygen at the rate required by thin polytetrafluoroethylene (PTFE), perfluoroalkoxy (PFA), and fluorinated ethylene propylene (FEP) membranes. Why, then, are highly permeable membranes most commonly chosen for oxygen electrodes? The reason is that, in some applications, speed of response and the ability to measure traces of dissolved oxygen are the most important considerations. Plus, the residual current is independent of the membrane chosen, whereas the sensitivity of the oxygen electrode is proportional to the oxygen permeability of the membrane. Hence, any sensor is more capable of measuring to lower concentration limits when it is fitted with a more permeable membrane than when fitted with a less permeable one.

A compromise between speed of response and high sensitivity, on the one hand, and stability and low stirring requirement on the other is therefore necessary when selecting a sensor for any particular application. When rapid response is indispensable, frequent recalibration is inevitable, and you will have to deal with some instability. When stability is of prime importance, a sluggish response is inevitable.

The list below is a useful summary of oxygen electrode drift and noise problems:

- General drift
- Osmotic effects (downward)
- Release of membrane stress (downward)
- Leakage or evaporation of a gross excess of electrolyte (upward)

- Deposition of anode metal at edge of cathode causing increase of cathode area (upward)
- Membrane erosion (upward)
- Membrane clogging (downward)
- Anode blocking (downward)
- Dilution of electrolyte with sample water (downward)
- Cathode deactivation by anode metal or poisons (downward)
- Short-term fluctuations
- Poorly stabilized anode supply voltage
- Poorly compensated temperature variations
- Flow rate (stirring) variations of sample
- Gravitational (orientation) effects
- Applied pressure effects
- Vibrational effects
- Intermediate-term fluctuations
- Barometric pressure effects
- Zero (residual) current variations.

I conclude this Appendix with two practical how-to sections: rechloriding the anode and polishing the cathode of Clark electrodes.

RECHLORIDING THE ANODE

Normally, the anode of an oxygen electrode is coated with a dark-brown layer of silver chloride (the brown color results from exposure to light; if the electrode has been kept in darkness, the color may be more of a fluorescent beige). If the coating is damaged, the electrode will not function correctly. It will drift and become noisy. Here I describe how to clean and rechloride the anode.

First, you will need the following:

- Analytical reagent grade 0.1 N HCl.
- A flashlight battery, A through C size, fresh, preferably in a battery holder for ease of connection. If you can't locate a suite battery holder, it's possible to solder wires directly to the battery if you have a high-wattage soldering iron. If you do this, be sure to heat the metal to which the solder will adhere and to allow the solder to flow onto it; don't just melt the solder, push it around, and hope it will adhere to the metal! Test your connection by trying to pull it off the battery; if it pulls off, do it again.
- A couple of inches of pure silver wire, which you can buy from a scientific supplier or (more cheaply) from a jeweler as "fine silver" wire.
- A crocodile clip or other sort of lead, to connect the battery or battery holder to the outside of the plug on the oxygen electrode cable.
- A beaker containing a couple of inches (4–5 cm) of the 0.1 N HCl.

Clip the silver wire to the side of the beaker so that it dips into the HCl on the inside. Connect the battery so that the silver wire is connected to the positive terminal of the battery and the ground shell of its plug is connected to the negative terminal. Remove the electrode's membrane, flush out the electrode with distilled water, and lower the electrode into the acid solution about a centimeter away from the silver wire. Let the acid solution creep up into the electrode until it just covers the anode. Watch carefully! Shine a strong light on the electrode so that you can see what is happening. You should see tiny hydrogen bubbles forming on the anode, which should slowly brighten as its chloride layer is stripped off until it appears silvery.

Now you want to replace the old chloride layer with a new one. To do this, just reverse the battery. Slowly, the anode will become light tan and then dark brown (remember that light "browns" the silver chloride, which is another reason for the bright light source you're shining on the anode). When the anode is a satisfactory shade of brown, similar to the color when new, remove the electrode from the beaker, rinse off the electrode, replace its membrane, and replace its electrolyte.

POLISHING THE CATHODE

In normal use, silver may be slowly deposited around or on the cathode of an oxygen electrode. If this happens, the cathode area increases, and with it the current that the electrode delivers. Various reactions can also occur at the cathode, causing instability or worse. If you suspect that the cathode of your oxygen electrode requires cleaning, the procedure is quite simple.

Using analytical reagent grade 6 M HNO_3, make 70 percent HNO_3, which is a standard commercial concentration. Dilute the acid 38 mL to 100 mL total with distilled water.

Remove the electrode's membrane. Rinse the electrode thoroughly with distilled water and gently scrub the cathode area clean of any obvious deposits. Now put a drop of the 6 M HNO_3 on a glazed porcelain surface or a microscope slide, *being very careful not to allow the acid anywhere near the anode, or it will dissolve*; touch just the tip of the electrode to the HNO_3 drop. Keep it there for exactly 2 min, then flush off the HNO_3 under a faucet, rinse the electrode with distilled water, replace its membrane, and replace its electrolyte. Some electrode designs might not allow this method to be used without risking the anode.

As an alternative, depending on your electrode's design, you may be able to use very, very fine (polishing-grade) sandpaper to clean the cathode. Do not use ordinary sandpaper; you should only use the finest grade of polishing paper, similar to the type used for polishing glass fiber-optic connections. Get this sandpaper from a jeweler or electronic supplier. Wet the sandpaper and very gently polish the tip of the probe, using a circular motion. This rather drastic method might be required if the cathode becomes slightly recessed. If the design of your probe causes the anode to be sanded as well (which is a better alternative to dissolving it), you will need to rechloride it as described above.

APPENDIX 3
DOWNLOADING THE PYTHON ENVIRONMENT

The following code is generic and should work with any scientific Python 3.x distribution in any operating system except for CP/M.

The scientific libraries numpy and matplotlib are required. They will be included in any scientifically oriented Python distribution. Google "scientific Python distribution" for the latest distributors, but here are three current versions worth considering:

- www.enthought.com/product/canopy/
- www.anaconda.com/
- If you run Windows I can highly recommend http://winpython.github.io/

The code below, which I have kept as simple as possible, is available for download from www.respirometry.org.

SIMULATING TIME CONSTANTS

```python
import numpy as np
import matplotlib.pyplot as plt

def lin_interp(array_in, finsize):
    '''
    takes array_in and returns it as an array
    with linear interpolation to finsize points
    '''
    out = np.zeros(finsize)
    max = len(array_in)
    step = float(finsize/max)
    index = 0
    while index < max-1:
        start = array_in[index]
        out[int(index * step)] = start
        inc = (array_in[index + 1] - array_in[index]) / step
        st = int(index * step) + 1
        en = int((index + 1) * step)
        for n in range(st, en):
            out[n] = start + (n - index * step) * inc
        index += 1
    return out
```

```
Change = 900 #change EE at this interval (seconds)
resting_value = 0.3
active_value = 0.8
dP = resting_value
dT = resting_value
time_constant = 240 #four minutes
i = 0 #elapsed time in seconds
kP = 1 / time_constant
kT = 1 / (time_constant * 5) #twenty minutes TC
p = resting_value

#set number of data points in simulation
maxi = 15000
#create the arrays to hold the simulated data
dp = np.zeros(maxi)
dl = np.zeros(maxi)
real_ee = np.zeros(maxi)
timebase = np.zeros(maxi)
#do the simulation
while i < maxi-1:
  i = i + 1
  j = int(i / Change) % 3
  if (j == 0):
      p = active_value
  else:
      p = resting_value
  dP = dP + (p - dP) * kP
  dT = dT + (p - dT) * kT
  timebase[i] = i
  real_ee[i] = p
  dp[i] = dP
  dl[i] = dT

#convert timebase from sec to min
timebase /= 60
'''
simulate multiplexing the 4 minute time constant
data at a 2.5 minute (150 sec) cycle time
'''
muxInt = 150
maxm = int(maxi / muxInt)
mdp = np.zeros(maxm)
mreal_ee = np.zeros(maxm)
mtimebase = np.zeros(maxm)
for n in range(0, maxm):
  j = n * muxInt
  mdp[n] = dp[j]
```

```python
  mreal_ee[n] = real_ee[j]
  mtimebase[n] = timebase[j]
mux=lin_interp(mdp,maxi)
#plot the real data
plt.figure(1)
ax1 = plt.subplot(311)
plt.plot(timebase, real_ee, 'k-')
plt.axis([10, 240, 0.2, 1])
plt.ylabel('Actual EE')
axes = plt.gca()
axes.set_ylim(0, 1.000001)
plt.setp(ax1.get_xticklabels(), visible = False)
plt.text(50, 0.05, "REAL EE")
#plot the 4 min TC data MUXed at 2.5 min
ax2 = plt.subplot(312, sharex = ax1)
plt.plot(timebase, mux, 'k-')
plt.ylabel('EE 4 min TC')
plt.axis([10, 240, 0.2, 1])
axes = plt.gca()
axes.set_ylim(0, 1.000001)
plt.setp(ax2.get_xticklabels(), visible = False)
plt.text(50, 0.05, "EE, 4 MINUTE TC, SAMPLED EVERY 2.5 MINUTES")
#plot the 20 min TC data second by second
plt.subplot(313, sharex = ax1)
plt.plot(timebase, dl, 'k-')
plt.ylabel('EE 20 min TC')
plt.axis([10, 240, 0.2, 1])
plt.xlabel('TIME (minutes)')
plt.text(50, 0.05, "EE, 20 MINUTE TC SAMPLED EVERY SECOND")
axes = plt.gca()
axes.set_ylim(0, 1.000001)

#tidy up the graph
plt.subplots_adjust(hspace=.2)
#save the graph and then display it
plt.savefig('Fig. 13.4.png', dpi = 600)
plt.show()
```

BIBLIOGRAPHY

Abreu-Vieira, G., Xiao, C., Gavrilova, O., and Reitman, M.L. (2015). Integration of body temperature into the analysis of energy expenditure in the mouse. *Mol Metab* 4: 461–70.

ACGIH. (1991). *Documentation of the Threshold Limit Values and Biological Exposure Indices*, 6th ed. Cincinnati, OH: American Conference of Governmental Industrial Hygienists, Inc.

Altmann, J. (1974). Observational study of behavior: sampling methods. *Behaviour* 49: 227–67.

Apte, M.G., Fisk, W.J., and Daisey, J.M. (2000). Associations between indoor CO_2 concentrations and sick building syndrome symptoms in US office buildings: an analysis of the 1994–1996 BASE Study Data (LBNL 44385). *Indoor Air* 10: 246–57.

Arch, J.R.S., Hislop D., Wang, S.J.Y., and Speakman, J.R. (2006). Some mathematical and technical issues in the measurement and interpretation of open-circuit indirect calorimetry in small animals. *Int J Obesity* 30: 1322–31.

Bakken, G.S. (1991). Time-resolved respirometry: equations for the simultaneous measurement of all respiratory gases and the calibration of oxygen consumption using variable inert gas flow rates. *J Therm Biol* 16: 313–15.

Bakken, G.S. (1992). Measurement and application of operative and standard operative temperatures in ecology. *Am Zool* 32: 194–216.

Bakken, G.S., Santee, W.R., and Erskine, D.J. (1985). Operative and standard operative temperature: tools for thermal energetics studies. *Am Zool* 25: 933–43.

Barcroft, J. and Haldane, J.S. (1902). A method of estimating the oxygen and carbonic acid in small quantities of blood. *J Physiol* 28: 232–40.

Bartholomew, G.A. (1958). The role of physiology in the distribution of terrestrial vertebrates. In: C.L. Hubbs, ed. *Zoogeography*. Publication 51, pp. 81–95. Washington, DC: American Association for the Advancement of Science.

Bartholomew, G.A. and Lighton, J.R.B. (1986). Oxygen consumption during hover-feeding in free-ranging Anna hummingbirds. *J Exp Biol* 123: 191–9.

Bartholomew, G.A., Vleck, D., and Vleck, C.M. (1981). Instantaneous measurements of oxygen consumption during pre-flight warm-up and post-flight cooling in sphingid and saturniid moths. *J Exp Biol* 90: 17–32.

Bartholomew, G.A., Lighton, J.R.B., and Louw, G.N. (1985). Energetics of locomotion and patterns of respiration in tenebrionid beetles from the Namib desert. *J Comp Physiol* 155: 155–62.

Bernstein, M.H., Hudson, D.M., Stearns, J.M., and Hoyt, R.W. (1977). Measurement of evaporative water loss in small animals by dew-point hygrometry. *J Appl Physiol* 43: 382–5.

Blanc, S., Géloën, A., Pachiaudi, C., Gharib, C., and Normand, S. (2000). Validation of the doubly labeled water method in rats during isolation and simulated weightlessness. *Am J Physiol* 279: 1964–79.

Breuner, C.W., Sprague, R.S., Patterson, S.H., and Woods, H.A. (2013). Environment, behavior and physiology: do birds use barometric pressure to predict storms? *J Exp Biol* 216: 1982–90.

Brodie, T.G. (1910). Some new forms of apparatus for the analysis of the gases of the blood by the chemical method. *J Physiol* 39: 391–6.

Brychta, R.J., Rothney, M.P., Skarulis, M.C., and Chen, K.Y. (2009). Optimizing energy expenditure detection in human metabolic chambers. *IEEE Eng Med Biol Soc* 2009: 6864–8. doi: 10.1109/IEMBS.2009.5333121.

Brychta, R., Wohlers, E., Moon, J., and Kong, C. (2010). Energy expenditure: measurement of human metabolism. *IEEE Eng Med Biol Soc* 29(1): 42–7. doi: 10.1109/MEMB.2009.935463.

Buck, A.L. (1981). New equations for computing vapor pressure and enhancement factor. *J Appl Meteorol* 20: 1527–32.

Burger, M. and van Breukelen, F. (2013). Construction of a low cost and highly sensitive direct heat calorimeter suitable for estimating metabolic rate in small animals. *J Thermal Biol* 38: 508–12.

Burnett, C.M.L. and Grobe, J.L. (2013). Direct calorimetry identifies deficiencies in respirometry for the determination of resting metabolic rate in C57Bl/6 and FVB mice. *Am J Physiol Endocrinol Metab* 305: E916–24.

Burnett, C.M.L. and Grobe, J.L. (2014). Dietary effects on resting metabolic rate in C57Bl/6 mice are differentially detected by indirect (O_2/CO_2 respirometry) and direct calorimetry. *Molec Metab* 3: 460–4.

Butler, A.A. and Kozak, L.P. (2010). A recurring problem with the analysis of energy expenditure in genetic models expressing lean and obese phenotypes. *Diabetes* 59: 323–9.

Carlson, S. (1996). Detecting micron-size movements. *Sci Am* 275: 96–9.

Catmull, E. and Rom, R. (1974). A class of local interpolating splines. In: R. Barnhill and R. Risenfeld, eds. *Computer Aided Geometric Design*, pp. 317–26. New York: Academic Press.

Choi, S.J., Yablonka-Reuveni, Z., Kaiyala, K.J., Ogimoto, K., and Schwartz, M.W. (2011). Increased energy expenditure and leptin sensitivity account for low fat mass in myostatin-deficient mice. *Am J Physiol Endocrinol Metab* 300: E1031–7.

Clark, L.C., Wolf, R., Granger, D., and Taylor, Z. (1953). Continuous recording of blood oxygen tensions by polarography. *J Appl Physiol* 6: 189–93.

Cooper, C.E. and Withers, P.C. (2010). Effect of sampling regime on estimation of basal metabolic rate and standard evaporative water loss using flow-through respirometry. *Physiol Biochem Zool* 83: 385–93.

Depocas, F. and Hart, J.S. (1957). Use of the Pauling oxygen analyzer for measurement of oxygen consumption of animals in open-circuit systems and in a short-lag, closed-circuit apparatus. *J Appl Physiol* 10: 388–92.

Djawdan, M., Rose, M.R., and Bradley, T.J. (1997). Does selection for stress resistance lower metabolic rate? *Ecology* 78: 828–37.

Domenech, T., Rafecas, I., Esteve, M., Argiles, J.M., and Alemany, M. (1988). A sensitive direct calorimeter for small mammals. *J Biochem Biophys Meth* 17: 35–42.

Einstein, A. (1933). *On the Method of Theoretical Physics*. New York: Oxford University Press.

Eliot, T.S. (1917). *Prufrock and Other Observations*. London: The Egotist.

Ellacott, K.L., Morton, G.J., Woods, S.C., Tso, P., and Schwartz, M.W. (2010). Assessment of feeding behavior in laboratory mice. Cell Metabol 12: 10–17.

Epting, R. J. (1980). Functional dependence of the power for hovering on wing disc loading in hummingbirds. *Physiol Zool* 53: 347–57.

Erdmann, C.A., Steiner, K.C., and Apte, M.G. (2002). Indoor carbon dioxide concentrations and sick building syndrome symptoms in the BASE Study Revisited: analyses of the 100 Building Dataset. *Proceedings of the 9th International Conference on Indoor Air Quality and Climate*, Monterey, Vol 3, pp. 443–8.

Fedak, M.A., Rome, L., and Seeherman, H.J. (1981). One-step N_2-dilution technique for calibrating open-circuit VO$_2$ measuring systems. *J Appl Physiol* 51: 772–6.

Frappell, P.B., Blevin, H.A., and Baudinette, R.V. (1989). Understanding respirometry chambers: what goes in must come out. *J Theoret Biol* 138: 479–94.

Garland, T., Jr, Dickerman, A.W., Janis, C.M., and Jones, J.A. (1993). Phylogenetic analysis of covariance by computer simulation. *Syst Biol* 42: 265–92.

Gibbs, A.G., Fukuzato, F., and Matzkin, L.M. (2003). Evolution of water conservation mechanisms in *Drosophila*. *J Exp Biol* 206: 1183–92.

Gilson, W.E. (1963). Differential respirometer of simplified and improved design. *Science* 141: 531–2.

Goldstein, G., Sharifi, M.R., Kohorn, L.U., Lighton, J.R.B., Shultz, L., and Rundel, P.W. (1991). Photosynthesis by inflated pods of a desert shrub, *Isomeris arborea*. *Oecologia* 85: 396–402.

Gordon, C.J., Becker, P., and Alia, J.S. (1998). Behavioral thermoregulatory responses of single- and group-housed mice. *Physiol Behav* 6: 255–62.

Groom, J.E., Toledo, C.B., Powers, D.R., Tobalske, B.W., and Welch, K.C. (2018). Integrating morphology and kinematics in the scaling of hummingbird hovering metabolic rate and efficiency. *Proc Biol Sci* 285: 20172011

Hamins, A.H., Bundy, M., and Dillon, S.E. (2005). Characterization of candle flames. *J Fire Protection Eng* 15: 265–85.

Hand, S.C. and Gnaiger, E. (1988). Anaerobic dormancy quantified in *Artemia* embryos: a calorimetric test of the control mechanism. *Science* 239: 1425–7.

Harter, T.S., Brauner, C.J., and Matthews, P.G.D. (2017). A novel technique for the precise measurement of CO_2 production rate in small aquatic organisms as validated on *Aeshnid* dragonfly nymphs. *J Exp Biol* 2017: 150235. doi: 10.1242/jeb.150235.

Hayes, J.P., Speakman, J.R., and Racey, P.A. (1992). Sampling bias in respirometry. *Physiol Zool* 65: 604–19.

Heinrich, B. (1993). *The Hot-Blooded Insects: Strategies and Mechanisms of Thermoregulation*. Cambridge, MA: Harvard University Press.

Held, N.M., Kuipers, E.N., van Weeghel, M., et al. (2018). Pyruvate dehydrogenase complex plays a central role in brown adipocyte energy expenditure and fuel utilization during short-term beta-adrenergic activation. *Sci Rep* 8: 9562

Henning, B., Lofgren, R. and Sjostrom, L. (1996). Chamber for indirect calorimetry with improved transient response. *Med Biol Eng Comput* 34: 207–12.

Heusner, A.A. (1991). Size and power in mammals. *J Exp Biol* 160: 25–54.

Heusner, A.A., Hurley, J.P., and Arbogas, T.R. (1982). Coulometric microrespirometry. *Am J Physiol* 243: 185–92.

Hoegh-Guldberg, O. and Manahan, D. (1995). Coulometric measurement of oxygen consumption during development of marine invertebrate embryos and larvae. *J Exp Biol* 198: 19–30.

Horowitz, P. and Hill, W. (2015). *The Art of Electronics*, 3rd ed. Cambridge: Cambridge University Press.

Jõgar, K., Kuusik, A., Metspalu, L., Hiiesaar, K., Luik, A., and Grishakova, M. (2005). Results of treatments with natural insecticidal substances on the development and physiological state of insects. *Agron Res* 4: 203–10.

Johnson, P.O. and Neyman, J. (1936). Tests of certain linear hypotheses and their application to some educational problems. *Stat Res Mem* 1: 57–93.

Johnson, T.A., Mercer, R.R., Taylor, P.C., Graham, J.A., and O'Neil, J.J. (1982). Oxygen consumption measured with microcomputer-assisted Warburg manometry. *J Appl Physiol* 53: 1634–7.

Kaiyala, K.J. and Ramsay, D.S. (2011). Direct animal calorimetry, the underused gold standard for quantifying the fire of life. *Comp Biochem Physiol A* 158: 252–64.

Kaiyala, K.J. and Schwartz, M.J. (2011). Toward a more complete (and less controversial) understanding of energy expenditure and its role in obesity pathogenesis. *Diabetes* 60: 17–23.

Kaiyala, K.J., Wisse, B., and Lighton, J.R.B. (2019). Validation of an equation for energy expenditure that does not require the respiratory quotient. In press, PLoS One.

Kaiyala, K.J., Morton, G.J., Leroux, B.G., Ogimoto, K., Wisse, B., and Schwartz, M.W. (2010). Identification of body fat mass as a major determinant of metabolic rate in mice. *Diabetes* 59: 1657–66.

Kleiber, M. (1961). *The Fire of Life*. New York: John Wiley & Sons.

Koteja, P. (1996). Measuring energy metabolism with open-flow respirometric systems: which design to choose? *Funct Ecol* 10: 675–7.

Kovacich, R.P., Martin, N.A., Clift, M.G., Stocks, C., Gaskin, I., and Hobby, J. (2006). Highly accurate measurement of oxygen using a paramagnetic gas sensor. *Measure Sci Technol* 17: 1579–85.

Lamprecht, I., Seymour, R.S., and Schultze-Motel, P. (1998). Direct and indirect calorimetry of thermogenic flowers of the sacred lotus, *Nelumbo nucifera*. *Thermochim Acta* 309: 5–16.

Lark, D.S., Kwan, J.R., McClatchey, P.M., et al. (2018). Reduced nonexercise activity attenuates negative energy balance in mice engaged in voluntary exercise. *Diabetes* 67(5): 831–40.

Lasiewski, R.C. (1963). Oxygen consumption of torpid, resting, active and flying hummingbirds. *Physiol Zool* 36: 122–40.

Levine, J.A. (2005). Measurement of energy expenditure. *Pub Health Nutr* 8: 1123–32.

Levy, A. (1964). The accuracy of the bubble meter method for gas flow measurements. *J Sci Instrum* 41: 449–53.

Lifson, N., Gordon, G.V., and McClintock, R. (1955). Measurement of total carbon dioxide production by means of D^2O^{18}. *J Appl Physiol* 7: 704–10.

Lighton, J.R.B. (1988). A simple, sensitive and versatile solid-state pressure transducer. *J Exp Biol* 134: 429–33.

Lighton, J.R.B. (1996). Discontinuous gas exchange in insects. *Annu Rev Entomol* 41: 309–24.

Lighton, J.R.B. (2007). Hot hypoxic flies: whole-organism interactions between hypoxic and thermal stressors in *Drosophila melanogaster*. *J Thermal Biol* 32: 134–43.

Lighton, J.R.B. (2008). *Measuring Metabolic Rates: A Manual for Scientists*. New York: Oxford University Press.

Lighton, J.R.B. (2012). "Instantaneous" metabolic measurement. *J Exp Biol* 215: 1605–6.

Lighton, J.R.B. (2015). Metabolic measurement techniques: baselining, mathematical correction of water vapour dilution and response correction. In: W. Gerrits and E. Labussiere, eds. *Indirect Calorimetry: Techniques, Computations and Applications*, pp. 57–72. Wageningen: Wageningen Academic Publishers.

Lighton, J.R.B. (2017). Limitations and requirements for measuring metabolic rates: a mini review. *Eur J Clin Nutr* 71: 301–5.

Lighton, J.R.B. and Bartholomew, G.A. (1988). Standard energy metabolism of a desert harvester ant, *Pogonomyrmex rugosus*: effects of humidity, temperature, body mass and group size. *Proc Natl Acad Sci USA* 85: 4765–9.

Lighton, J.R.B. and Duncan, F.D.D. (1995). Standard and exercise metabolism and the dynamics of gas exchange in the giant red velvet mite, *Dinothrombium magnificum*. *J Insect Physiol* 41: 877–84.

Lighton, J.R.B. and Duncan, F.D.D. (2002). Energy cost of locomotion: validation of laboratory data by in situ respirometry. *Ecology* 83: 3517–22.

Lighton, J.R.B. and Duncan, F.D.D. (2005). Shaken, not stirred: a serendipitous study of ants and earthquakes. *J Exp Biol* 208: 3103–7.

Lighton, J.R.B. and Feener, D.H. (1989). Running in a desert ant: a comparison of energetics and ventilation during voluntary and forced locomotion. *Nature* 342: 174–5.

Lighton, J.R.B. and Halsey, L.G. (2011). Flow-through respirometry applied to chamber systems: pros and cons, hints and tips. *Comp Biochem Physiol* 158: 265–75.

Lighton, J.R.B. and Schilman, P.E. (2007). Oxygen reperfusion damage in an insect. *PLoS One* 2(12): e1267.

Lighton, J.R.B. and Turner, R.J. (2004). Thermolimit respirometry: an objective assessment of critical thermal maxima in two sympatric desert harvester ants, *Pogonomyrmex rugosus* and *P. californicus*. *J Exp Biol* 207: 1903–13.

Lighton, J.R.B., Bartholomew, G.A., and Feener, D.H. (1987). Energetics of locomotion and load carriage and a model of the energy cost of foraging in the leaf-cutting ant *Atta colombica*. *Physiol Zool* 60: 524–37.

Lighton, J.R.B., Brownell, P.H., Joos, B., and Turner, R.J. (2001). Low metabolic rate in scorpions: implications for population biomass and cannibalism. *J Exp Biol* 204: 607–13.

Longo, K.A., Charoenthongtrakul, S., Giuliana, D.J., et al. (2010). The 24-hour respiratory quotient predicts energy intake and changes in body mass. *Am J Physiol Regul Integr Comp Physiol* 298: R747–54.

Lusk, G. (1928). *The Elements of the Science of Nutrition*, 4th ed. London: W.B. Saunders.

MacKay, S. J., Loiseau, A., Poivre, R., and Huot, A. (1991). Calibration method for small animal indirect calorimeters. *Am J Physiol Endocrinol Metab* 261: 661–4.

Mata, A.J., Caloin, M., Robin, J.-P., and Le Maho, Y. (2006). Reliability in estimates of body composition of birds: oxygen-18 versus deuterium dilution. *Physiol Biochem Zool* 79: 202–9.

Melanson, E.L., Ingebrigtsen, J.P., Bergouignan, A.,Ohkawara, K., Kohrt, W.M., and Lighton, J.R.B. (2010). A new approach for flow-through respirometry measurements in humans. *Am J Physiol* 298: 1571–9.

Meyer, C.W., Klingenspor, M., Rozman, J., and Heldmaier, G. (2004). Gene or size: metabolic rate and body temperature in obese growth hormone-deficient dwarf mice. *Obes Res* 12: 1509–18. doi: 10.1038/oby.2004.188.

Meyer, C.W., Reitmeir, P., and Tschöp, M.H. (2015). Exploration of energy metabolism in the mouse using indirect calorimetry: measurement of daily energy expenditure (DEE) and basal metabolic rate (BMR). *Curr Protocol Mouse Biol* 5: 205–22.

Moon, J.K., Vohra, F.A., Valerio Jimenez, O.S., Puyau, M.R., and Butte, N.F. (1995). Closed-loop control of carbon dioxide concentration and pressure improves response of room respiration calorimeters. *J Nutr* 125: 220–8.

Nagy, K.A. (1989). Doubly-labelled water studies of vertebrate physiological ecology. In: P.W. Rundel, J.R. Ehleringer, and K.A. Nagy, eds. *Stable Isotopes in Ecological Research*, pp. 270–87. New York: Springer.

Nguyen, T., De Jonge, L., Smith, S.R., and Bray, G.A. (2003). Chamber for indirect calorimetry with accurate measurement and time discrimination of metabolic plateaus of over 20 min. *Med Biol Engineer Comput* 41: 572–8.

Pfeiffer, T.J., Summerfelt, S.T., and Watten, B.J. (2011). Comparative performance of CO_2 measuring methods: marine aquaculture recirculation system application. *Aquacult Engineer* 44: 1–9.

Prentice, D.A. and Miller, D.T. (1996). Pluralistic ignorance and the perpetuation of social norms by unwitting actors. *Adv Exp Social Psychol* 28: 161–209.

Ravussin, E., Lillioja, S., Anderson, T.E., Christin, L., and Bogardus, C. (1986). Determinants of 24-hour energy expenditure in man: methods and results using a respiratory chamber. *J Clin Invest* 78: 1568–78.

Rezende, E.L., Bozinovic, F., and Garland, T. Jr. (2004). Climatic adaptation and the evolution of basal and maximum rates of metabolism in rodents. *Evolution* 58: 1361–74.

Satish, U., Mendell, M.J., Shekhar, K., Hotchi, T., Sullivan, D., and Streufert, S. (2012). Is CO_2 an indoor pollutant? Direct effects of low-to-moderate CO_2 concentrations on human decision-making performance. *Environ Health Perspect* 120(12): 1671–7.

Savitzky, A. and Golay, M.J.E. (1964). Smoothing and differentiation of data by simplified least squares procedures. *Anal Chem* 36: 1627–39.

Sazanov, E.S. and Schuckers, S. (2009). The energetics of obesity: a review. Monitoring energy uptake and energy expenditure in humans. *IEEE Eng Med Biol* 29(1): 31–5. doi: 10.1109/MEMB.2009.935470.

Schilder, R.J. and Marden, J.H. (2006). Metabolic syndrome and obesity in an insect. *Proc Natl Acad Sci USA* 103: 18805–9.

Schoffelen, P.F.M., Westerterp, K.R., Wim, H.M., Saris, W.H.M., and Ten Hoor, F. (1997). A dual-respiration chamber system with automated calibration. *J Appl Physiol* 83: 2064–72.

Shannon, C.E. and Weaver, W. (1963). *The Mathematical Theory of Communication*. Champaign: University of Illinois Press.

Singh, B.N. and Mathur, P.B. (1936). A manometer for comparative study of physiological processes. *Biochem J* 30: 323–5.

Sláma, K. (1988). A new look at insect respiration. *Biol Bull* 175: 289–300.

Speakman, J.R. (1998). The history and theory of the doubly labeled water technique. *Am J Clin Nutr* 68: 932–8.

Speakman, J.R. (2013). Measuring energy metabolism in the mouse: theoretical, practical, and analytical considerations. *Frontiers Physiol* 4: 34.

Speakman, J.R. (2014). Should we abandon indirect calorimetry as a tool to diagnose energy expenditure? Not yet. Perhaps not ever. Commentary on Burnett and Grobe (2014). *Molec Metabol* 4: 342–4.

Speakman, J.R. and Racey, P.A. (1986). Measurement of CO_2 production by the doubly labeled water technique. *J Appl Physiol* 61: 1200–2.

Stephens, B.B., Bakwin, R.M., Tans, P.P., Teclaw, R.M., and Baumann, D.D. (2007). Application of a differential fuel-cell analyzer for measuring atmospheric oxygen variations. *J Atmos Oceanic Technol* 24: 82–94. doi: 10.1175/JTECH1959.1.

Suarez, R.K., Lighton, J.R.B., Moyes, C.D., Brown, G.S., Gass, C.L., and Hockachka, P.W. (1990). Fuel selection in rufous hummingbirds: ecological implications of metabolic biochemistry. *Proc Natl Acad Sci USA* 87: 9207–10.

Sun, M., Reed, G.W., and Hill, J.O. (1994). Modification of a whole room indirect calorimeter for measurement of rapid changes in energy expenditure. *J Appl Physiol* 76: 2686–91.

Symonds, M.R.E. and Elgar, M.A. (2002). Phylogeny affects estimation of metabolic scaling in mammals. *Evolution* 56: 2330–3.

Tanner, J.M. (1949). Fallacy of per-weight and per-surface area standards, and their relation to spurious correlation. *J Appl Physiol* 2: 1–15.

Tartes, U., Kuusik, A., and Vanatoa, A. (1999). Diversity in gas exchange and muscular activity patterns in insects studied by a respirometer-actograph. *Physiol Entomol* 24: 150–7.

Taylor, C., Heglund, N.C., and Maloiy, G.M. (1982). Energetics and mechanics of terrestrial locomotion. I. Metabolic energy consumption as a function of speed and body size in birds and mammals. *J Exp Biol* 97: 1–21.

Taylor, P. (1977). A continuously recording respirometer, used to measure oxygen consumption and estimate locomotor activity in tsetse flies, *Glossina morsitans*. *Physiol Entomol* 12: 45–50.

Tohjima, Y., Machida, T., Tomonori, W., Akam, I., Amara, T., and Moriwak, Y. (2005). Preparation of gravimetric standards for measurement of atmospheric oxygen and reevaluation of atmospheric oxygen concentration. *J Geophys Res* 110, D11302.

Tschöp, M.H., Speakman, J.R., Arch, J.R., et al. (2012). A guide to analysis of mouse energy metabolism. *Nat Methods* 9: 57–63.

Van Klinken, J.B., van den Berg, S.A., Havekes, L.M., and Van Dijk, K.W. (2012). Estimation of activity related energy expenditure and resting metabolic rate in freely moving mice from indirect calorimetry data. *PloS One* 7(5): e36162.

Van Voorhies, W., Melvin, R.G., Ballard, J.W.O., and Williams, J.B. (2008). Validation of manometric microrespirometers for measuring oxygen consumption in small arthropods. *J Insect Physiol* 54: 1132–7.

Vioque, J., Ramos, R.M., Navarrete-Muñoz, E.M., and García-de-la-Hera, M. (2010). A bibliometric study of scientific literature on obesity research in PubMed (1988–2007). *Obes Revs* 1: 603–61.

Vleck, D. (1987). Measurement of O_2 consumption, CO_2 production, and water vapor production in a closed system. *J Appl Physiol* 62: 2103–6.

Walsberg, G.E. and Hoffman, T.C.M. (2005). Direct calorimetry reveals large errors in respirometric estimates of energy expenditure. *J Exp Biol* 208: 1035–43.

Weir, J.B. (1949). New methods of calculating metabolic rate with special reference to protein metabolism. *J Physiol* 109: 1–9.

Welch, K.C. and Suarez, R.K. (2007). Oxidation rate and turnover of ingested sugar in hovering Anna's (*Calypte anna*) and rufous (*Selasphorus rufus*) hummingbirds. *J Exp Biol* 210: 2154–62.

Werthessen, N. T. (1937). An apparatus for the measurement of the metabolic rate of small animals. *J Biol Chem* 119: 233–9.

White, C.R. (2003). Allometric analysis beyond heterogeneous regression slopes: use of the Johnson–Neyman technique in comparative biology. *Phys Biochem Zool* 76: 135–40.

Whitman, W. (1855). *Leaves of Grass*. Ed. M. Cowley, 1986. New York: Penguin.

Wickler, S.J., Hoyt, D.F., Cogger, E.A., and Myers, G. (2003). The energetics of the trot–gallop transition. *J Exp Biol* 206: 1557–64.

Willard, H.H. and Smith, G.F. (1922). The preparation and properties of magnesium perchlorate and its use as a drying agent. *J Am Chem Soc* 44: 2255–9.

Withers, P.C. (2001). Design, calibration and calculation for flow-through respirometry systems. *Austral J Zool* 49: 445–61.

Wu, L. and He, H. (1994). Preparation of perlite-based magnesium perchlorate desiccant with colour indicator. *Chem Educ* 41: 633–7.

Xu, T. (2007). Characterization of mini environments in a clean room: design characteristics and environmental performance. *Build Environ* 42: 2993–3000.

INDEX